Transition Metal Catalyzed Pyrimidine, Pyrazine, Pyridazine and Triazine Synthesis

Transition Metal Catalyzed Pyrimidine, Pyrazine, Pyridazine and Triazine Synthesis

Transition Metal-Catalyzed Heterocycle Synthesis Series

Xiao-Feng Wu and Zechao Wang

Leibniz-Institut für Katalyse e. V., Rostock, Germany

ELSEVIER

AMSTERDAM • BOSTON • HEIDELBERG • LONDON
NEW YORK • OXFORD • PARIS • SAN DIEGO
SAN FRANCISCO • SINGAPORE • SYDNEY • TOKYO

Elsevier
Radarweg 29, PO Box 211, 1000 AE Amsterdam, Netherlands
The Boulevard, Langford Lane, Kidlington, Oxford OX5 1GB, United Kingdom
50 Hampshire Street, 5th Floor, Cambridge, MA 02139, United States

Notices
Knowledge and best practice in this field are constantly changing. As new research and experience broaden our
understanding, changes in research methods, professional practices, or medical treatment may become
necessary.

Practitioners and researchers must always rely on their own experience and knowledge in evaluating and using
any information, methods, compounds, or experiments described herein. In using such information or methods
they should be mindful of their own safety and the safety of others, including parties for whom they have a
professional responsibility.

British Library Cataloguing-in-Publication Data
A catalogue record for this book is available from the British Library

Library of Congress Cataloging-in-Publication Data
A catalog record for this book is available from the Library of Congress

ISBN: 978-0-12-809378-8

For Information on all Elsevier publications
visit our website at https://www.elsevier.com

Working together
to grow libraries in
developing countries

www.elsevier.com • www.bookaid.org

Publisher: John Fedor
Acquisition Editor: Katey Birtcher
Editorial Project Manager: Jill Cetel
Production Project Manager: Vijayaraj Purushothaman
Designer: MPS

Typeset by MPS Limited, Chennai, India

CONTENTS

Contents

CHAPTER 1

Introduction

Pyrimidine is a class of heterocyclic aromatic organic compounds, containing 2 nitrogen atoms at positions 1 and 3 of the 6-member ring. In nature, the pyrimidine ring is synthesized from glutamine, bicarbonate, and aspartate [1]. Pyrimidine as an integral part of DNA and RNA, demonstrates a diverse array of biological and pharmacological activities, including acting as an antitubercular, antimicrobial [2], antiplatelet [3], and antifungal agent [4], as well as having antiviral properties [5] (Scheme 1.1). Therefore, the synthesis of pyrimidines has become much more important. During the past years, various methodologies on pyrimidines synthesis have been developed quickly [6]. In this volume, the synthesis of pyrimidine will be elaborated.

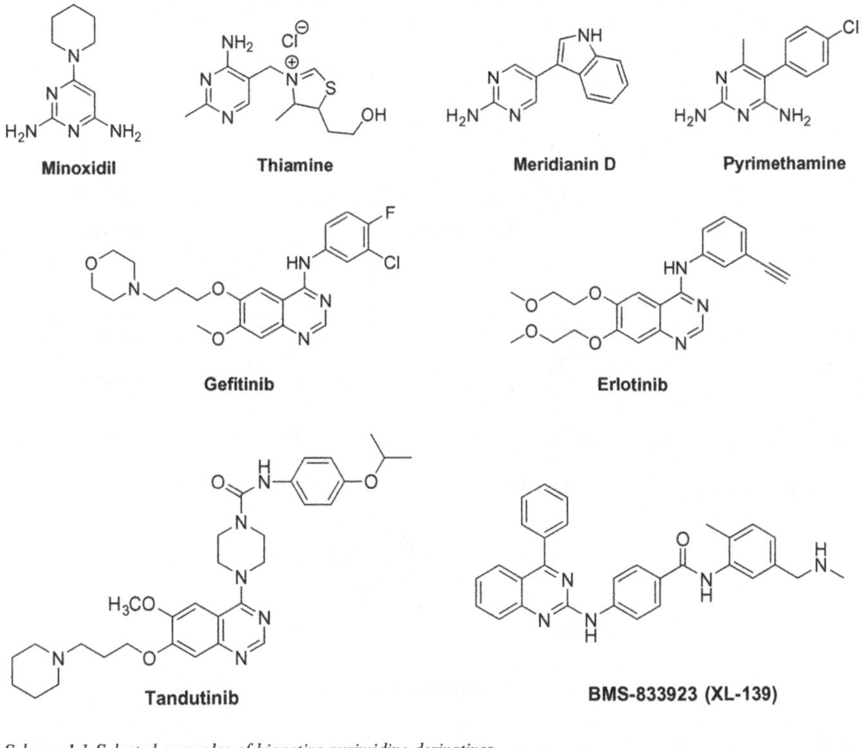

| Minoxidil | Thiamine | Meridianin D | Pyrimethamine |

| Gefitinib | Erlotinib |

| Tandutinib | BMS-833923 (XL-139) |

Scheme 1.1 Selected examples of bioactive pyrimidine derivatives.

Transition Metal Catalyzed Pyrimidine, Pyrazine, Pyridazine and Triazine Synthesis.
DOI: http://dx.doi.org/10.1016/B978-0-12-809378-8.00001-8

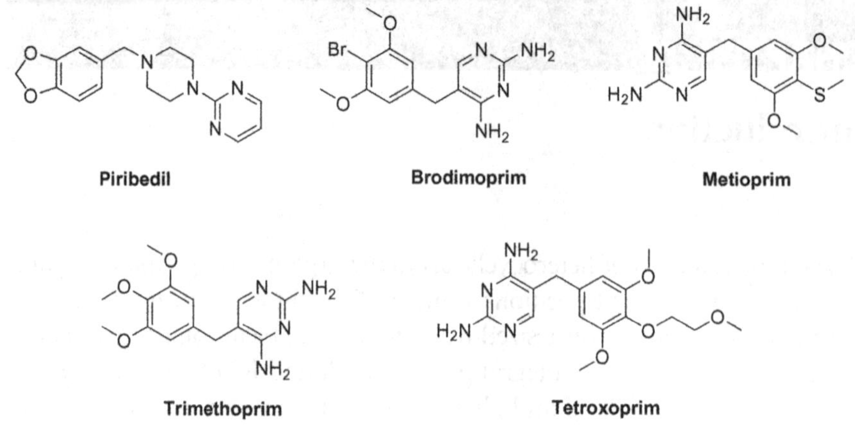

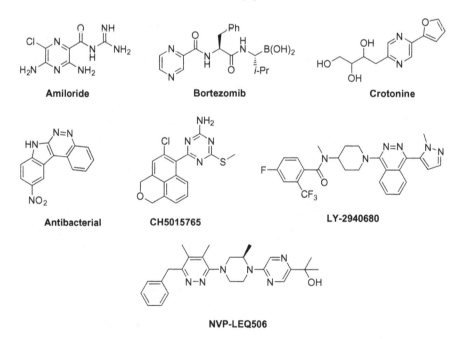

Scheme 1.1 (Continued)

Other related six-member heterocyclic compounds such as pyrazine, pyridazine, and triazine will be discussed here as well. They are also highly significant in the investigation for new medicines, active pharmaceutical ingredients (API), and fine chemicals (Scheme 1.2) [7]. Their syntheses are attractive as well [8].

Scheme 1.2 Selected examples of bioactive compounds.

REFERENCES

[1] Lagoja, I. M. *Chem. Biodiversity* **2005**, *2*, 1−50.

[2] Kamdar, N. R.; Haveliwala, D. D.; Mistry, P. T.; Patel, S. K. *Eur. J. Med. Chem.* **2010**, *45*, 5056−5063.

[3] Bruno, O.; Brullo, C.; Ranise, A.; Schenone, S.; Bondavalli, F.; Barocelli, E., et al. *Bioorg. Med. Chem. Lett.* **2001**, *11*, 1397−1400.

[4] El-Hashash, M. A.; Mahmoud, M. R.; Madboli, S. A. *Indian J. Chem.* **1993**, *32B*, 449−452.

[5] Shamroukh, A. H.; Zaki, M. E. A.; Morsy, E. M. H.; Abdel-Motti, F. M.; Abdel-Megeid, F. M. E. *Arch. Pharm.* **2007**, *340*, 236−243.

[6] a. Brown, D. J. The Pyrimidines. In *The Chemistry of Heterocyclic Compounds;* Weissberger, A., Ed.; Vol. 16; Wiley-Interscience: New York, 1970.
 b. Lister, J. H. Fused Pyrimidines, Part II, The Purines. In *The Chemistry of Heterocyclic Compounds;* Weissberger, A., Taylor, E. C., Eds.; Vol. 24; Wiley-Interscience: New York, 1971.
 c. Hurst, D. T. *An Introduction to the Chemistry and Biochemistry of Pyrimidines, Purines and Pteridines;* Wiley: Chichester, 1980.
 d. Brown, D. J. In *Comprehensive Heterocyclic Chemistry;* Katritzky, A. R., Rees, C. W., Eds.; Vol. 3; Pergamon Press: Oxford, 1984. Chap. 2.13.
 e. Bojarski, J. T.; Mokrosz, J. L.; Bartón, H. J.; Paluchowska, M. H. *Adv. Heterocycl. Chem* **1985**, *38*, 229−297.
 f. Hoffmann, M. G. In *Houben-Weyl, Methoden der organischen Chemie;* Schaumann, E., Ed.; Vol. E9; Thieme Verlag: Stuttgart, 1996.
 g. Hill, M. D.; Movassaghi, M. *Chem. Eur. J.* **2008**, *14*, 6836−6844.

[7] a. Sato, N. In Comprehensive Heterocyclic *Chemistry II;* Katritzky, A. R., Rees, C. W., Boulton, A. J., Eds.; Elsevier: Oxford, 1996.
 b. Sidwell, R. W.; Dixon, G. J.; Sellers, S. M.; Schabel, F. M. *Appl. Microbiol.* **1968**, *16*, 370−392.
 c. Falke, D.; Rada, B. *Acta Virol.* **1970**, *14*, 115−123.
 d. Klenke, B.; Stewart, M.; Barrett, M. P.; Brun, R.; Gilbert, I. H. *J. Med. Chem.* **2001**, *44*, 3440−3452.
 e. Sahoo, S.; Veliyath, S. K.; Kumar, M. C. B. *Int. J. Res. Pharm. Sci.* **2012**, *3*, 326−333.
 f. Plech, T.; Luszczki, J. J.; Wujec, M.; Flieger, J.; Pizon, M. *Eur. J. Med. Chem.* **2013**, *60*, 208−215.

[8] a. Okada, Y.; Taguchi, H.; Nishiyama, Y.; Yokoi, T. *Tetrahedron Lett* **1994**, *8*, 1231−1234.
 b. Chen, C.; DagninoJr, R.; McCarthy, J. R. *J. Org. Chem.* **1995**, *60*, 8428−8430.
 c. Zhang, C. Y.; Tour, J. M. *J. Am. Chem. Soc.* **1999**, *121*, 8783−8790.
 d. Brzozowskia, Z.; Saczewskia, F.; Gdaniecb, M. *Eur. J. Med. Chem.* **2000**, *35*, 1053−1064.
 e. Palacios, F.; Retana, A. M. O.; Gil, J. I.; Munain, R. L. *Org. Lett.* **2002**, *4*, 2405−2408.
 f. Liu, C.; Lin, J.; Leftheris, K.; Klenke, B.; Stewart, M.; Barrett, M. P., et al. *J. Med. Chem.* **2001**, *44*, 3440−3452.
 g. Pridmore, S. J.; Slatford, P. A.; Taylor, J. E.; Whittlesey, M. K.; Williams, J. M. J. *Tetrahedron* **2009**, *65*, 8981−8986.
 h. Kaila, J. C.; Baraiya, A. B.; Pandya, A. N.; Jalani, H. B.; Sudarsanam, V.; Vasu, K. K. *Tetrahedron Lett* **2010**, *51*, 1486−1489.
 i. Zhu, C.; Yamane, M. *Tetrahedron* **2011**, *67*, 4933−4938.

CHAPTER 2

Synthesis of Pyrimidine

2.1 SYNTHESIS OF PYRIMIDINE BY [4 + 2] CYCLIZATION REACTIONS

In 2005, Lejon and coworkers developed a palladium-catalyzed synthesis of pyrimidines (Scheme 2.1) [1]. In the presence of palladium (II) acetate and triphenylphosphine, by reacting α-methyl or α-methylene ketones with formamide, satisfactory yields of pyrimidines can be obtained by this procedure. No significant yield difference could be detected using ketones with a methyl- or a methylene group.

Scheme 2.1

To a round-bottomed flask charged with palladium(II) acetate (40 mg, 0.18 mmol) and triphenylphosphine (95 mg, 0.36 mmol) were added formamide (5.0 g, 110 mmol), iodobenzene (2.0 g, 10 mmol), and ketone (3.6 mmol) and the resulting mixture was heated at 160°C for 8 h. The reaction mixture was diluted with ether (50 mL) and extracted with three 20 mL portions of a 2 M aqueous solution of hydrogen chloride. The combined aqueous phase was basified by addition of solid sodium hydroxide or a 4 M aqueous solution of sodium hydroxide. The combined layers were extracted with ether (2 × 50 mL) and washed with water and brine before drying over sodium carbonate. The solvent was removed under vacuum and the crude products were purified by flash chromatography.

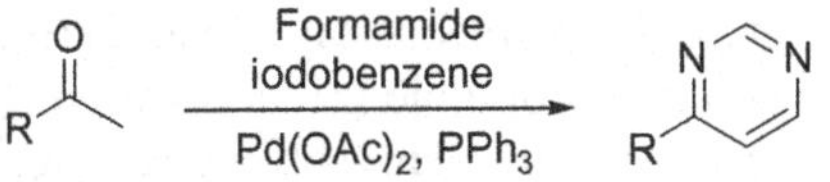

Scheme 2.1 Palladium-catalyzed synthesis of pyrimidines.

Transition Metal Catalyzed Pyrimidine, Pyrazine, Pyridazine and Triazine Synthesis.
DOI: http://dx.doi.org/10.1016/B978-0-12-809378-8.00002-X

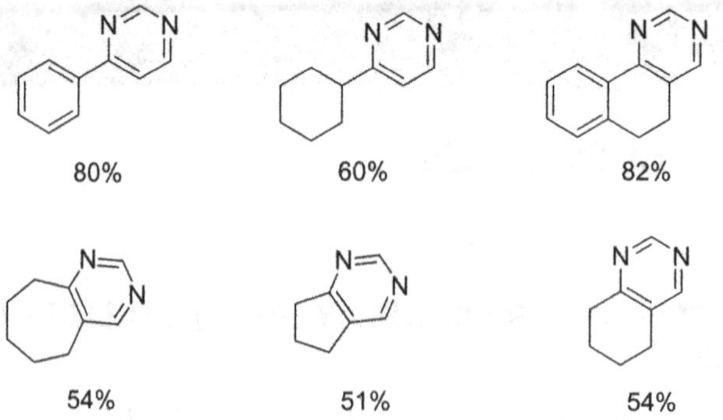

80%   60%   82%

54%   51%   54%

Scheme 2.1 (Continued)

In 2012, Yavari and coworkers reported a one-pot procedure for the synthesis of pyrimidines from sulfonyl azides, terminal alkynes, and cyanoguanidine. Moderate to good yields of pyrimidine derivatives were produced (Scheme 2.2) [2]. This method might be considered a practical route for the synthesis of functionalized sulfonamidopyrimidines. The advantages of this methodology include short reaction times and readily available starting materials and catalysts. Aromatic alkynes participated in the coupling to furnish the corresponding diaminopyrimidine derivatives in good yields. Aliphatic acetylenes served as lower yielding substrates compared to phenylacetylenes. Aromatic and aliphatic sulfonyl azides worked efficiently and the desired pyrimidine derivatives were obtained in good yields.

Scheme 2.2

To a mixture of azide (1.2 mmol), alkyne (1 mmol), CuI (0.1 mmol), and Et$_3$N (1 mmol) in THF (2 mL) was slowly added, under N$_2$ atmosphere, cyanoguanidine (1 mmol). The mixture was stirred at room temperature. After completion of the reaction (12 h), the mixture was diluted with CH$_2$Cl$_2$ (2 mL) and aqueous NH$_4$Cl solution (3 mL), stirred for 30 min, and the layers separated. The aqueous layer was extracted with CH$_2$Cl$_2$ (10 mL × 3) and the combined organic fractions dried (Na$_2$SO$_4$) and concentrated under reduced pressure. The solvent was removed under vacuum and the crude products were purified by flash chromatography.

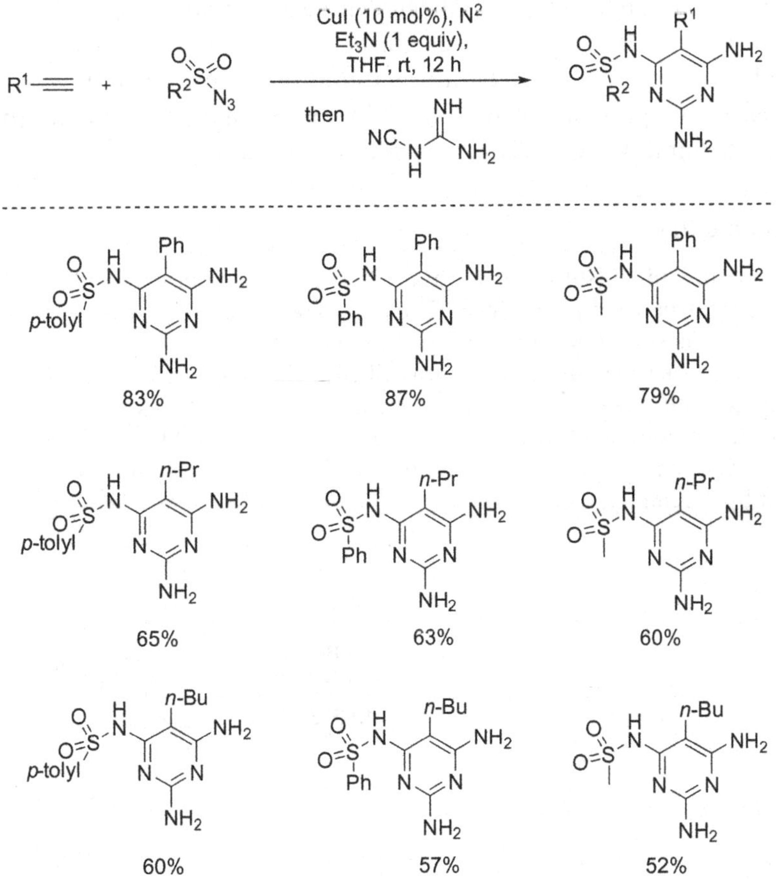

Scheme 2.2 Copper-catalyzed one-pot synthesis of pyrimidine.

In 2016, Shafiee and coworkers reported a simple and straightforward procedure for the synthesis of a series of polyfunctionalized pyrimidines. In the presence of CuBr/Et$_3$N in DMF (Dimethylformamide) at 80°C via the reaction between N-(substituted carbamothioyl) benzimidamides and malononitrile, good yields of the desired products were observed (Scheme 2.3) [3]. Various N-(substituted carbamothioyl) benzimidamides were reacted with malononitrile under the optimized conditions to obtain different polyfunctionalized pyrimidines. All substrates possessing electron-rich as well as electron-poor substituents underwent the cyclization reaction leading to the formation of the related products within 2 h in good yields. This new synthetic approach for polyfunctionalized pyrimidines synthesis would be beneficial for both organic and medicinal chemists to develop novel pyrimidine-based drugs [4]. Interestingly, it was found that if the reaction was conducted in the presence of molecular iodine, a different synthetic route proceeded and N-benzo[d]thiazole could be obtained.

Scheme 2.3

A mixture of N-(substituted carbamothioyl) benzimidamides (2 mmol), malononitrile (2 mmol), CuBr (2 mmol), and NEt₃ (8 mmol) in DMF (2 mL) was stirred at 80°C for 2 h. After completion of reaction (checked by TLC), the reaction mixture was filtered through a bed of celite and washed with 30 mL ethyl acetate. Then, water (30 mL) was added to the filtrate, the organic layer was extracted, dried over Na_2SO_4, and the solvent was removed under reduced pressure. The crude product was purified by column.

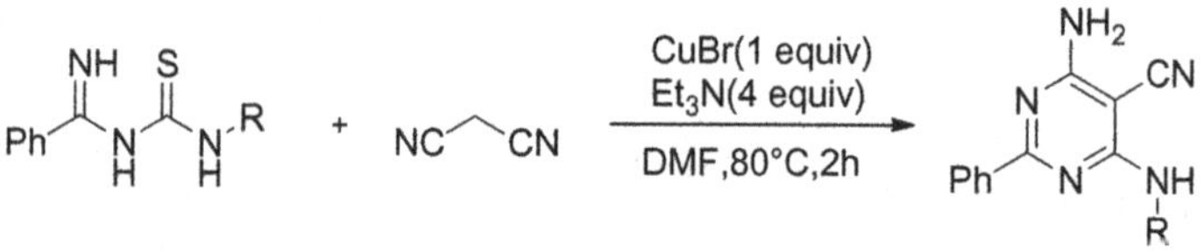

Scheme 2.3 Copper-catalyzed synthesis of polyfunctionalized pyrimidines.

2.2 SYNTHESIS OF PYRIMIDINE BY [3 + 2 + 1] CYCLIZATION REACTIONS

In 2003, Müller's group developed a novel method to synthesize pyrimidines, which used acid chlorides and terminal alkynes in the reaction under Sonogashira conditions (Scheme 2.4) [5]. It represents a straightforward one-pot three-component method to produce 2,4-di- and 2,4,6-trisubstituted pyrimidines in moderate to good yields according to a highly flexible coupling–addition–cyclocondensation sequence. In this reaction, only one equivalent of trimethylamine was used. This approach represented a consecutive combination of modern cross-coupling methodology and classic Michael addition cyclocondensation. Aromatic, heteroaromatic, and aliphatic substituted pyrimidines were synthesized by this procedure.

Scheme 2.4

In a screw cap pressure vessel Pd(PPh$_3$)$_2$Cl$_2$ (14.0 mg, 0.02 mmol) and CuI (7.0 mg, 0.04 mmol) were dissolved in degassed THF or MeCN (5 mL). Then Et$_3$N (0.14 mL, 1.00 mmol for alkyne or 0.17 mL, 1.25 mmol for all other alkynes) as well as acid chloride (1 mmol) and alkyne (1 mmol) were added successively to the solution. The reaction mixture was stirred for 1–3 h until the conversion was complete (monitored by TLC). Finally, Na$_2$CO$_3$ 10H$_2$O (973.0 mg, 3.40 mmol for alkyne or 687.0 mg, 2.40 mmol for all other alkynes 2) and amidinium chloride (1.2 mmol or 2.50 mmol of guanidine hydrochloride 6 g) were added to the suspension and the reaction mixture was heated to reflux temperature for 12–14 h. After cooling to r.t., the crude products were purified by chromatography on silica gel to give the analytically pure pyrimidines.

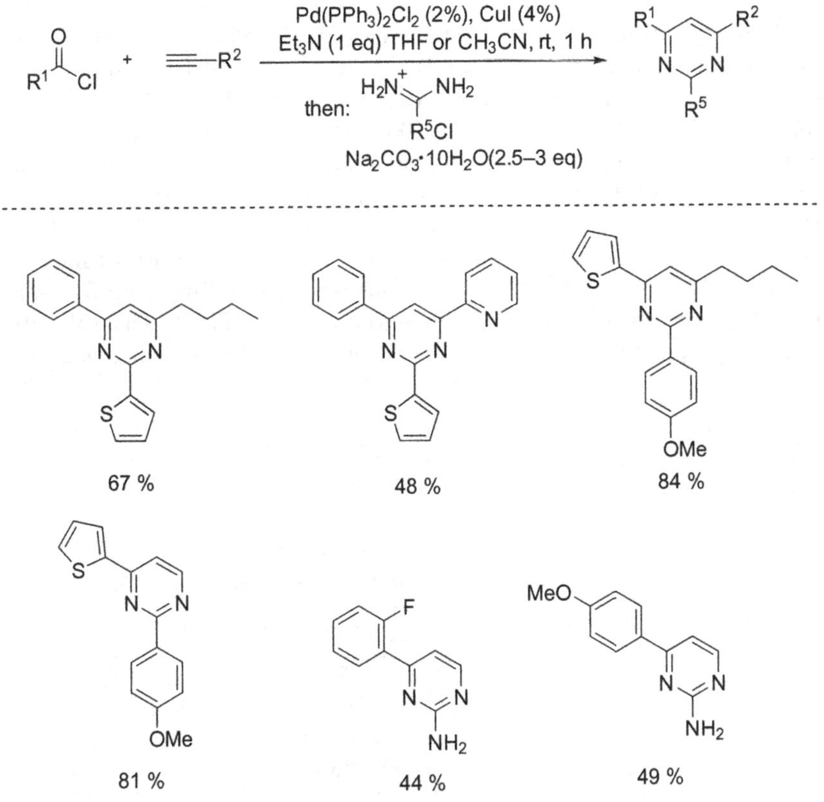

Scheme 2.4 *One-pot coupling–cyclocondensation synthesis of pyrimidines.*

In 2003, Müller and coworker developed a new one-pot, three-component synthesis of 2,4-substituted pyrimidines (Scheme 2.5) [6]. The crucial point for the successful Sonogashira coupling of (hetero) aroyl chlorides and (Trimethylsilyl)-acetylene was the application of only one equiv of triethylamine as the essential base. In addition, this new protocol represented a more atom-economical alternative to the state-of-the-art Stille synthesis of (TMS)-ynones. The mild reaction conditions (stoichiometric amount of triethylamine at room temperature) were particularly well suited for the development of novel multicomponent reactions. Applying these peculiar conditions to a variety of (hetero)aroyl chlorides, the corresponding (TMS)-alkynones were obtained in moderate to good yields. Sensitive or fragile functionalities such as nitro, bromo, or acetoxy substituents were tolerated well here.

Scheme 2.5

A stirred mixture of 14 mg (0.02 mmol) of $Pd(PPh_3)_2Cl_2$, and 7 mg (0.04 mmol) of CuI in 5 mL of CH_3CN was degassed for 5 min. Then 0.14 mL (1.00 mmol) of triethylamine, 147 mg (1.00 mmol) of 2-thienylacid chloride and 0.14 mL (1.00 mmol) of trimethylsilyl acetylene were added. The reaction mixture was stirred for 1 h under nitrogen at room temperature until the alkyne was completely consumed (monitored by TLC). Then a solution of 1.0 g (3.5 mmol) of $Na_2CO_3 \cdot 10\,H_2O$, 240 mg (2.50 mmol) of guanidinium hydrochloride, and 5 mL of methanol was added. The reaction mixture was heated to reflux temp for 14 h until the conversion was complete (monitored by TLC). The solvents were evaporated and the residue was chromatographed on silica gel.

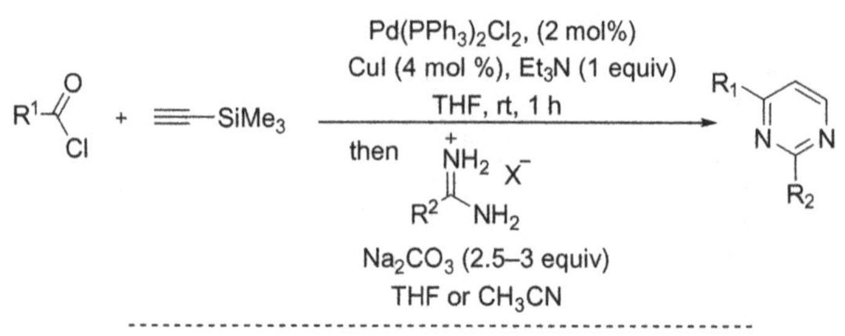

Scheme 2.5 *Palladium-catalyzed one-pot, three-component synthesis of pyrimidines.*

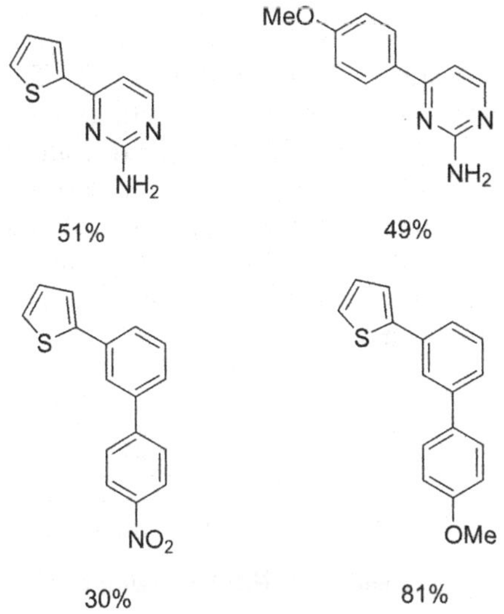

MeO

S

N N
NH₂

N N
NH₂

51%   49%

S

S

NO₂   OMe

30%   81%

Scheme 2.5 (Continued)

In 2005, Müller and coworkers developed one-pot four-component synthesis of pyrimidines by a carbonylative coupling–cyclocondensation
sequence. (Scheme 2.6) [7]. Reaction of (hetero)aryl iodides and terminal
alkynes in THF at room temperature under 1 atm of carbon monoxide
(a CO-filled balloon attached to the reaction vessel) and in the presence of
2 equiv of triethylamine and catalytic amounts of [Pd(PPh₃)₂Cl₂] and CuI
for 48 h followed by addition of the amidinium salts in the presence of 2.5
equiv of sodium carbonate in acetonitrile/water gave the 2,4,6-trisubsti-
tuted pyrimidines moderate to good yield. It was noteworthy that this
sequence proceeds efficiently only for neutral or electron-rich aryl iodides.
When electron-deficient aryl iodides were examined, a 1:2 mixture of carbonylative and noncarbonylative products was isolated. On the other
hand, attempts to replace the aryl iodides with aryl bromides to increase
the selectivity led to the complete loss of reactivity, and only the starting
material was recovered. This method could be considered as a complemen-
tary approach when acid-sensitive functionality could not be tolerated.

Scheme 2.6

In a Schlenk flask [Pd(PPh₃)₂Cl₂] (35 mg, 0.05 mmol), CuI (4 mg,
0.02 mmol), para-iodoanisole (234 mg, 1.00 mmol), and THF (5 mL)
were placed under nitrogen. Carbon monoxide was bubbled through the
solution for 5 min, and then hexyne (0.14 mL, 1.20 mmol) and triethylamine (0.28 mL, 2.00 mmol) were added successively. The reaction mixture

became dark red and was stirred at room temperature for 48 h under 1 atm of carbon monoxide (balloon filled with CO). Then the suspension was treated with Na_2CO_3 (265 mg, 2.50 mmol), amidinium salt (195 mg, 1.20 mmol), water (0.5 mL), and CH_3CN (5 mL), and the reaction mixture was heated at reflux for 24 h. After cooling to room temperature, the reaction mixture was diluted with brine (20 mL) and extracted with dichloromethane (5 × 20 mL). The combined organic layers were dried with sodium sulfate, concentrated to dryness, and subjected to column chromatography on silica gel.

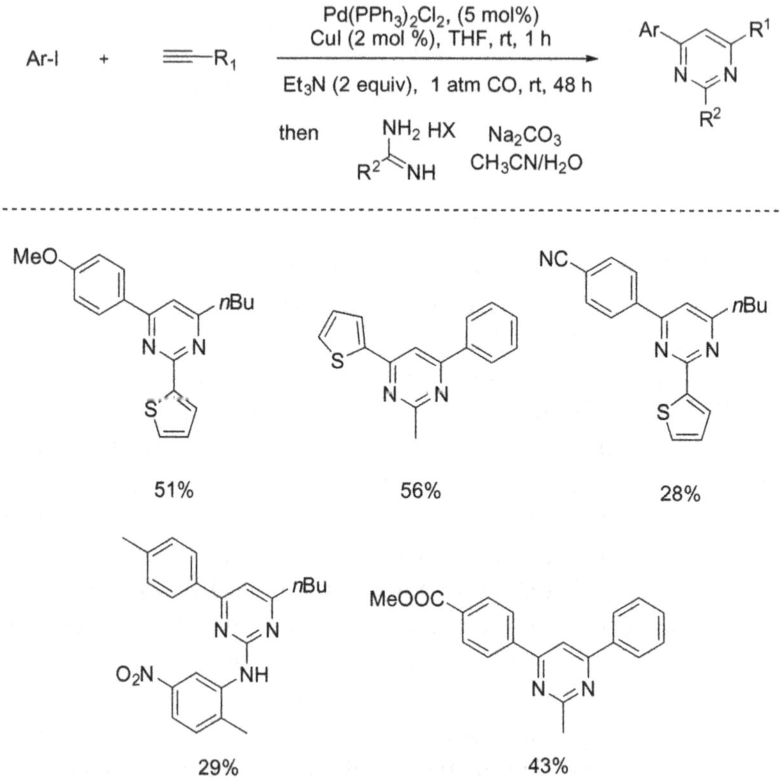

Scheme 2.6 One-pot four-component synthesis of pyrimidines.

In 2008, Stonehouse and coworkers described a palladium-catalyzed four-component reaction for the generation of pyrimidines, which was a one-pot process (Scheme 2.7) [8]. In this reaction, a wide range of aryl halides, terminal alkynes, molybdenum hexacarbonyl, and amidines were shown to be an efficient method for the construction of highly substituted pyrimidines. The efficiency of the process was reduced when the halogen was changed from iodine to bromine, with no reaction observed when chlorobenzene was used. Substitution of the aryl halide had little effect on reaction yield. While benzyl bromide was used, none of the product was detected. Various terminal alkynes worked well under this condition except ethyl propiolate, which gave no desired product. Most amidines were shown in moderate to good yields via this procedure. However, when they used urea in this reaction, no desired product was obtained. While changing the nucleophile to guanidine, they only got trace amounts of the desired product. And they concluded that urea and guanidine were not reactive enough to participate in the reaction sequence due to their ability to delocalize their electron density.

Scheme 2.7

In a 25-mL 3-neck round-bottomed flask were added halide (1.0 mmol), alkyne (1.5 mmol), molybdenum hexacarbonyl (1.5 mmol), nucleophile (2.0 mmol), palladium acetate (0.05 mmol), copper iodide (0.02 mmol), tri-*tert*-butylphosphine (0.1 mmol) and cesium carbonate (2.5 mmol, 5.0 mmol when nucleophile was used as a salt) in toluene (2.5 mL), and MeCN (2.5 mL) and the mixture was heated at 80°C overnight. The solvent was removed and the residue was dissolved in CH_2Cl_2 and filtered through celite. This crude material was purified by flash chromatography (gradient elution with PE and EtOAc) and then by reverse-phase preparative HPLC on a Xbridge C8 19×50 mm column (gradient elution with 95% of 0.3% aq NH_3 solution−5% MeCN to 5% of 0.3% aq NH_3 solution-95% MeCN).

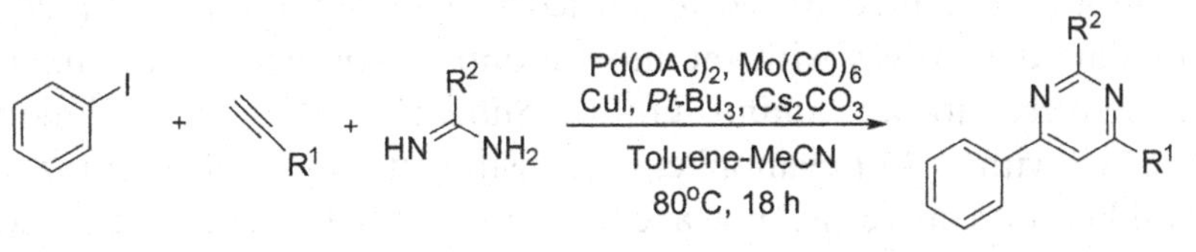

Scheme 2.7 One-pot four-component reaction for the synthesis of pyrimidines.

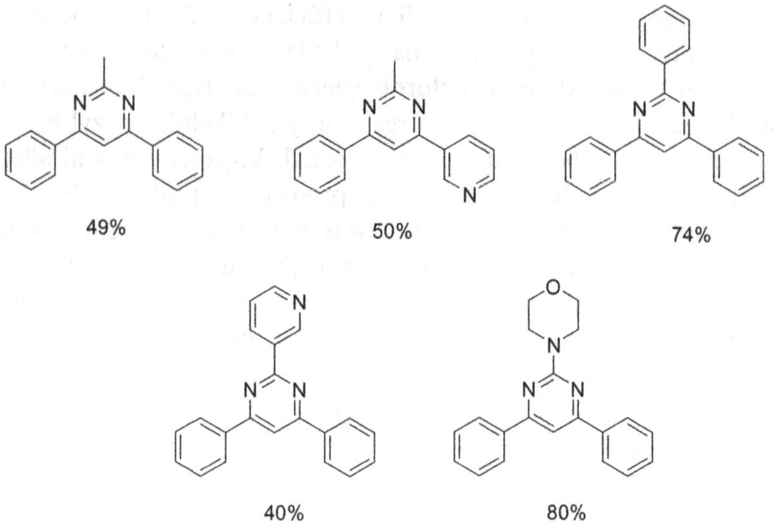

49%   50%   74%

40%   80%

Scheme 2.7 (Continued)

In 2013, Rostamizadeh and coworkers reported a one-pot, three-component reaction to synthesis pyrimidines. Substrates such as an aldehyde, malononitrile, and benzamidine hydrochloride were transformed in the presence of magnetic nano Fe_3O_4 particles as the catalyst under solvent-free conditions (Scheme 2.8A) [9]. The pyrimidine-5-carbonitrile derivatives in this article were all prepared with excellent yields. Both aromatic aldehydes with electron-donating substituents and electron-withdrawing substituents showed significant reactivity in this process. Then they used pyrimidine-5-carbonitrile derivatives to react with hydrazine hydrates to get more pyrimidine derivatives (Scheme 2.8B). When the pyrimidine derivatives with the nitro group were used, the reaction led to the formation of pyrazolo-pyrimidine derivatives with the nitro group on the phenyl ring having been reduced to an amino group. Similarly, when a pyrimidine derivative with a 4-(4-cyanophenyl) substituent was used, a pyrazolo-pyrimidine derivative lacking a cyano group on the phenyl ring was obtained. When bulky aryl substituents were placed at position 4 of the pyrimidine ring, the addition of hydrazine hydrate led to decomposition of the pyrimidine to the benzylidiene malononitrile and benzamidine intermediates rather than the formation of related pyrazolo[3,4-d] pyrimidine derivatives.

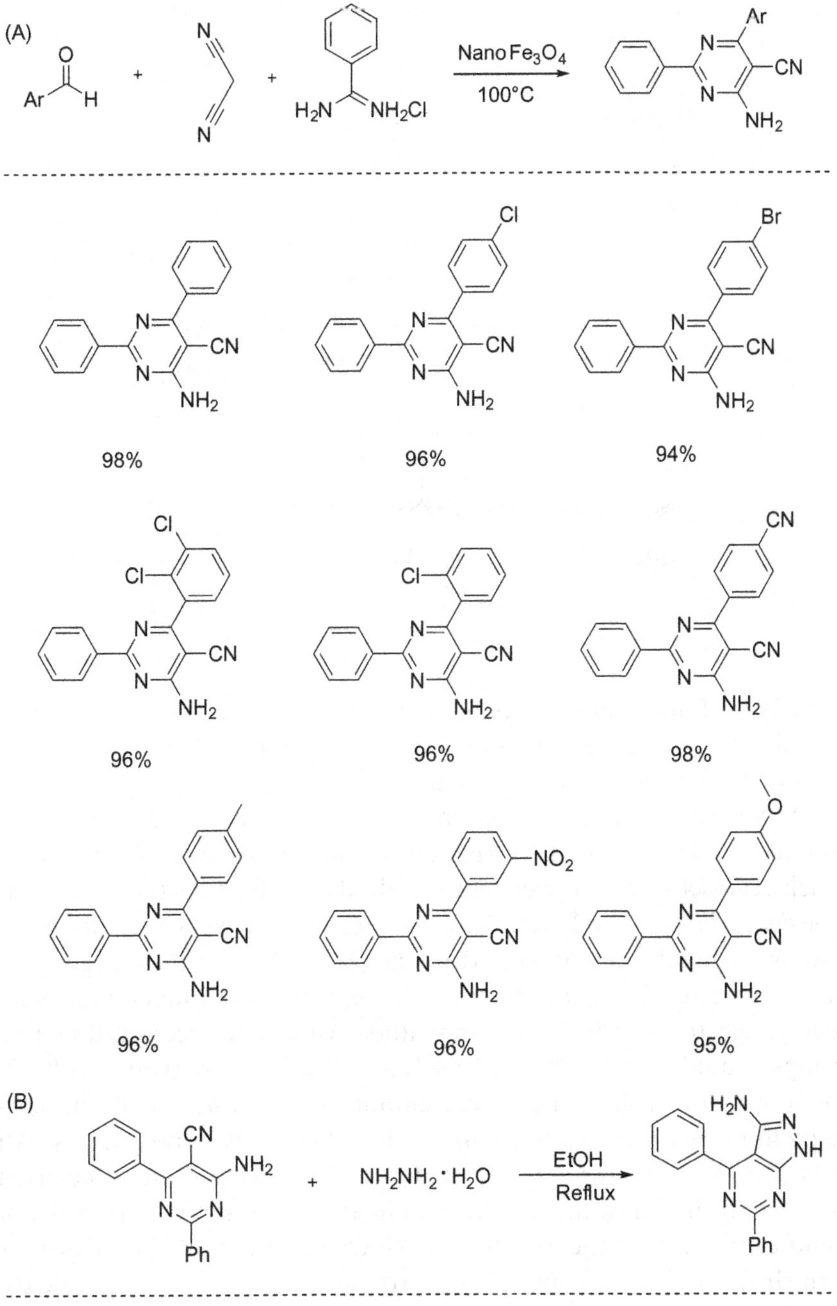

Scheme 2.8 *(A) Synthesis of pyrimidine-5-carbonitrile derivatives. (B) Further transformation of pyrimidine-5-carbonitriles.*

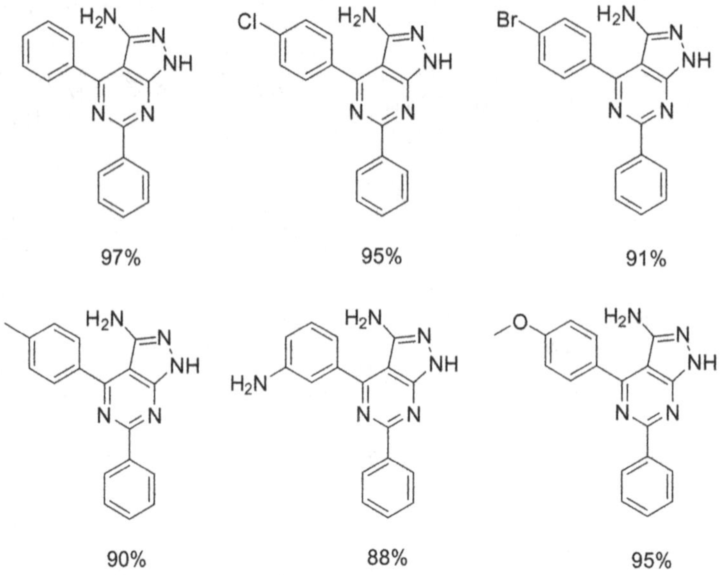

97%   95%   91%

90%   88%   95%

Scheme 2.8 (Continued)

In 2014, Kumar and coworkers developed an efficient regioselective cascade procedure for the synthesis of pyrimidine derivatives. This transformation proceeded via a transition-metal (copper/silver) catalyzed coupling reaction between 2-aminobenzimidazole, aldehydes, and alkynes leading to the formation of propargylamine intermediate, which regioselectively underwent 6-endo-dig cyclization through intramolecular N–H bond activation interceded C–N bond formation leading to highly functionalized imidazo[1,2-a]pyrimidines in good to excellent yields (Scheme 2.9) [10]. A wide range of aromatic aldehydes bearing electron-neutral, electron-withdrawing, and electron-donating groups could be employed as the coupling partners with 2-aminobenzimidazole, which were smoothly transformed to the corresponding pyrimidines with alkyne in good to moderate yields. All *para-*, *meta-*, and *ortho*-substituted aldehydes were easily converted into the desired products, which indicated that steric bulk did not significantly affect the reactivity. Aldehydes bearing dioxo groups furnished the corresponding products in moderate yields. Moreover, aliphatic aldehydes also reacted smoothly to give the desired coupling products in good yields. Unfortunately, aromatic nitrogen-containing aldehydes such as pyridine-4-carboxyaldehyde and *N,N*-dimethylbenzaldchydc failed to provide the desired coupling products. Aromatic alkynes bearing methyl groups at the *para*- and *meta*-position afforded the corresponding products in good yields. The tolerance of functional groups such as chloro, fluoro, bromo, nitro, methoxy, and cyano in this method provided an opportunity for further chemical transformations.

Scheme 2.9

2-Aminobenzimidazole (133.1 mg, 1.0 mmol), *p*-chloro-benzaldehyde (140 mg, 1.0 mmol), and phenylacetylene (0.220 mL, 2.0 mmol) was dissolved in 5 mL of acetonitrile in a round-bottom flask, CuI (38 mg, 20 mol%) was added, the mixture was stirred at room temperature for 10 min under nitrogen, silver carbonate (54 mg, 20 mol%) was added, and the reaction mixture was refluxed (81 − 83°C) until full consumption of 2-aminobenzimidazole, as monitored through TLC. Upon completion (6 − 8 h) of the reaction, the mixture was filtered on Celite. The filtrate was concentrated under reduced pressure to give the crude material, which was purified by column chromatography on silica gel (eluent: EtOAc/hexane) and afforded benzoimidazopyrimidine.

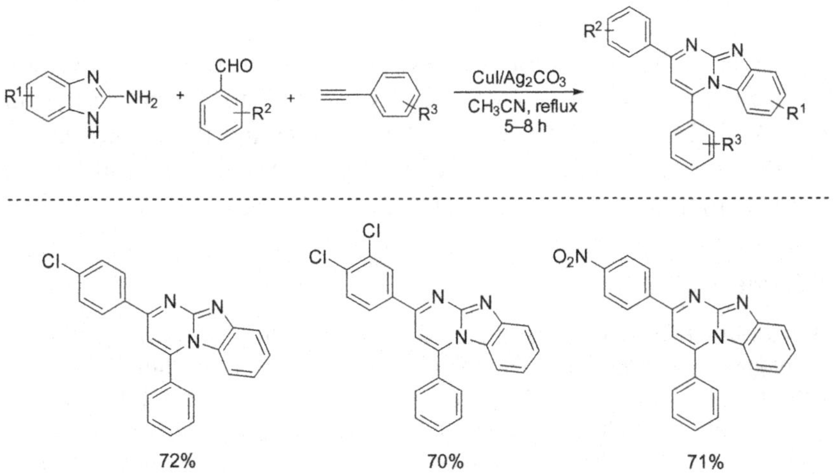

Scheme 2.9 Synthesis of imidazopyrimidine derivatives.

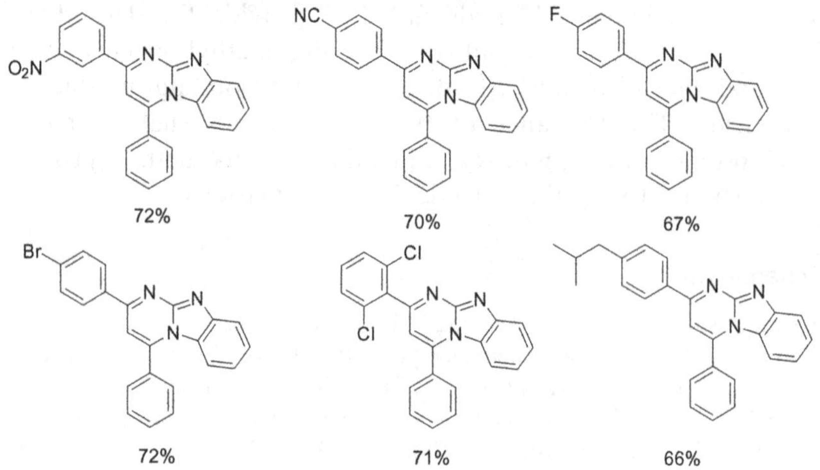

72%   70%   67%

72%   71%   66%

Scheme 2.9 (Continued)

In 2015, Kempe's group reported a novel, regioselective, iridium-catalyzed multicomponent synthesis of pyrimidines from amidines and alcohols, which proceeded via a sequence of condensation and dehydrogenation steps (Scheme 2.10) [11]. While the condensation steps deoxygenated the alcohol components, the dehydrogenations led to aromatization. Two equiv of hydrogen and water were liberated in the course of the reactions. PN5P − Ir − pincer complexes, catalyzed this sustainable multicomponent process most efficiently. In this reaction, many primary and secondary alcohol were used to synthesize pyrimidines. Aryl chlorides, heterocycles like pyridine and thiophene, as well as olefins were tolerated. A variety of unsymmetrically and fully substituted pyrimidines were synthesized in isolated yields of up to 93%.

Scheme 2.10

A pressure tube (Ace pressure tube, 38 mL volume) was charged with a magnetic stirring bar, alcohol (2.0 eq.), primary alcohol (2.2 eq.), *t*-BuOK (2.0 eq.) catalyst Ir (as a 0.01 M stock solution in 1,4-dioxane, 1.0−2.0 mL depending on catalyst loading) under an inert atmosphere (glove box). The tube was closed with a cap and immersed into a pre-heated oil bath (125°C). After 4 h the reaction was cooled and amidine (1.0 eq.), alcohol (1.1 eq.) and *t*-AmOH (2 mL) were added under inert atmosphere (glove box). The mixture was heated to reflux under inert atmosphere in an open system for 24 h. The reaction was cooled, quenched with water (1 mL), and extracted with ethyl acetate

$(3 \times 20$ mL). The crude reaction mixture was analyzed by GC-MS and after drying and evaporation of solvent the remainder was subjected to flash column chromatography on silica gel.

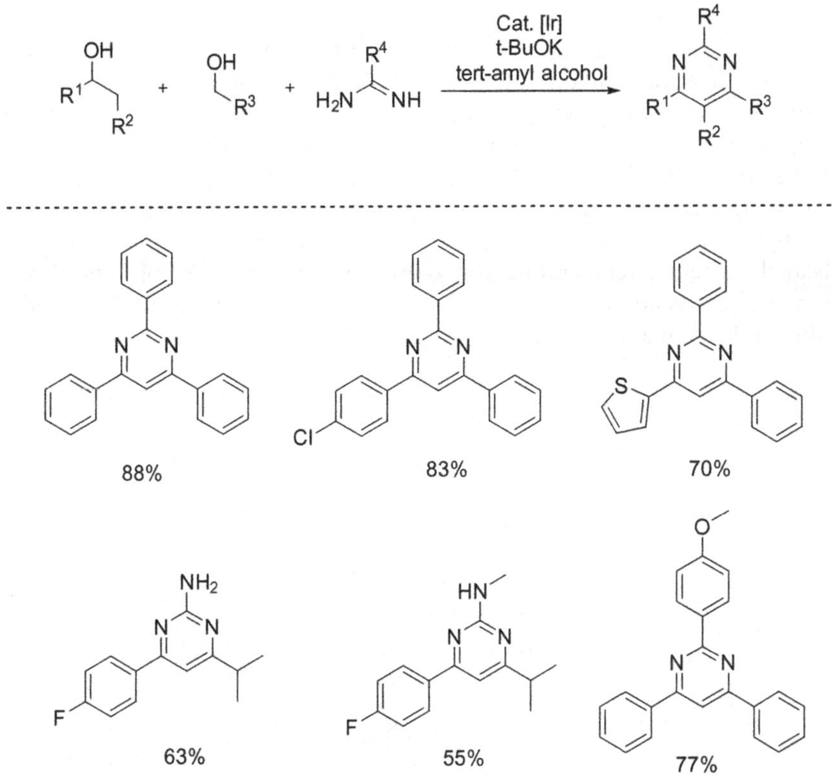

Scheme 2.10 A sustainable multicomponent pyrimidine synthesis.

2.3 SYNTHESIS OF PYRIMIDINE BY [3 + 3] CYCLIZATION REACTIONS

In 2000, a three-component one-pot pyrimidine synthesis based upon a coupling–isomerization sequence was developed by Müller and coworkers (Scheme 2.11) [12]. p-Iodo nitrobenzene or 4-bromo pyridine, several aryl propynols, and amidinium salts were reacted under the reaction conditions of the Sonogashira coupling in a boiling mixture of triethylamine and THF. Here, the electron-withdrawing nature of the (hetero)aryl halide was crucial for the successful coupling–isomerization step in this reaction. The substituents on the pyrimidine ring could be electron-rich, electron-poor, or heterocyclic and even bromo substituents were tolerated.

Scheme 2.11

To a magnetically stirred solution of 0.25 g (1.00 mmol) of 4-iodo nitrobenzene, 22 mg (0.02 mmol) of Pd(PPh₃)₂Cl₂, and 2 mg (0.01 mmol) of CuI in a degassed mixture of 10 mL of THF and 5 mL of triethylamine under nitrogen was added a solution of 145 mg (1.05 mmol) of 1-(3-thienyl)-propyn-1-ol in 10 mL of THF dropwise at room temperature over a period of 30 min. The mixture was heated to reflux temperature for 16 h. After the mixture cooled to room temperature, 165 mg (1.00 mmol) of 2-thienyl amidinium chloride was added and the reaction mixture was heated to reflux temperature for 48 h. After cooling the solvents were removed in vacuo, and the residue was dissolved in dichloromethane and filtered through a short pad of silica gel.

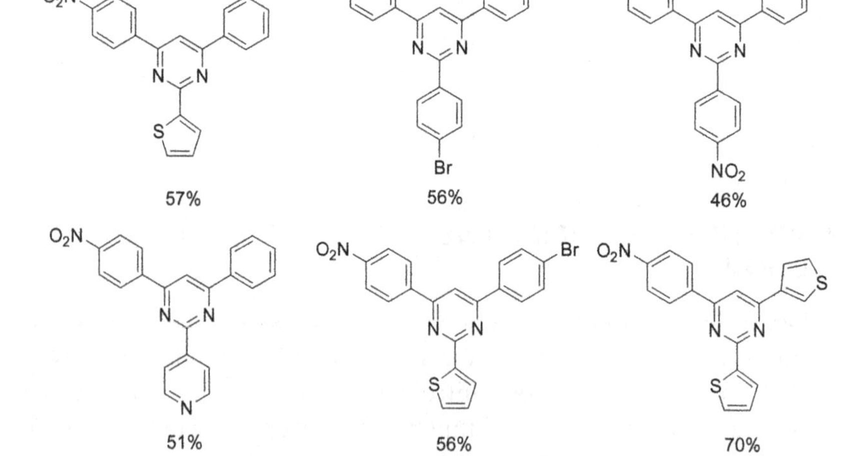

Scheme 2.11 Palladium-catalyzed one-pot synthesis of pyrimidines.

In 2006, Langer's group reported the synthesis of 4-(3-hydroxyalkyl)pyrimidines by ring transformation reactions of 2-alkylidenetetrahydrofurans with amidines [13]. This methodology relied on the condensation of amidines with 2-alkylidene-tetrahydrofuran derivatives. These substituted tetrahydrofuran substrates were prepared from one-pot cyclization reactions of free and masked 1,3-dicarbonyl dianions and gave access to various 2,4,6-trisubstituted pyrimidine derivatives.

In 2007, Hu and coworkers developed an efficient method to generate a diversified pyrimidine library via a sequential one-pot reaction of iodochromone, arylboronic acid, and amidine by Suzuki coupling and condensation (Scheme 2.12) [14]. The electronic variations on both the aryl group of the boronic acid and the substitution of amidine gave the desired product in moderate to good yield. The compounds with 4-methoxyphenyl or CH_3 were assayed for the inhibition of human hepatocellular carcinoma cell line BEL-7402. According to the bioassay result, they synthesized several derivatives using different substituted iodochromones. However, all these substituents, including OMe, Cl, NO_2, and CH_3, at the paraposition of OH resulted in less inhibition against the growth of BEL-7402 cells.

Scheme 2.12

A mixture of iodochromone (1.2 mmol) and aryl boronic acids (1.1 equiv) in the presence of 2% $Pd(PPh_3)_4$ and 2.0 equiv K_2CO_3 in 5 mL THF-H_2O (4:1) was refluxed overnight and then split into six portions, to which was added 1.5 equiv of amidines (0.3 mmol) and 1,8-diazabicyclo [5.4.0]undec-7-ene (DBU; 0.3 mmol or 0.6 mmol) for each portion. The mixture was stirred at 50−60°C for about 10 h, and the corresponding products were obtained by flash chromatography.

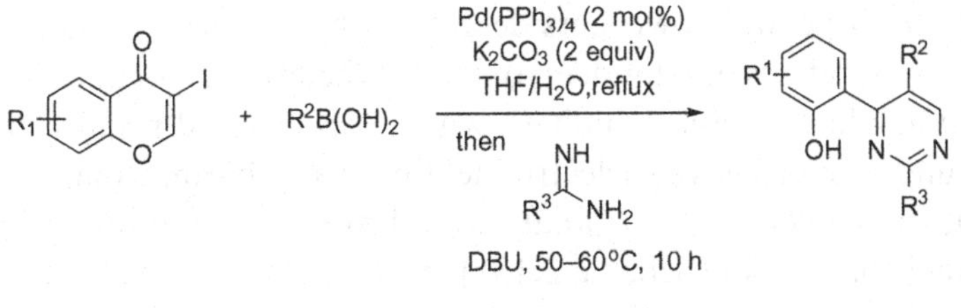

Scheme 2.12 One-pot synthesis of 2,4,5-substituted pyrimidines.

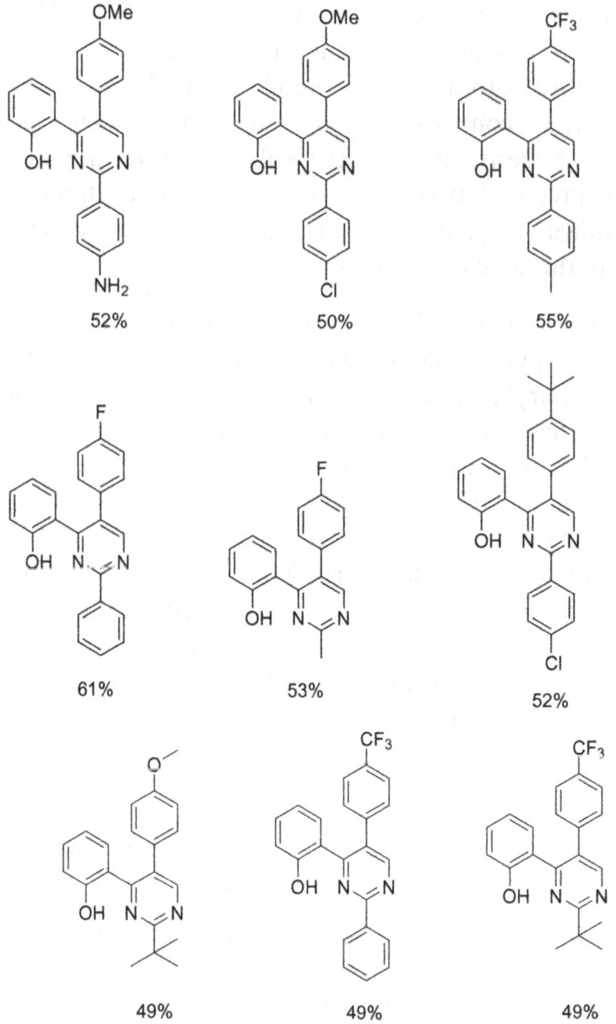

52%   50%   55%

61%   53%   52%

49%   49%   49%

Scheme 2.12 (Continued)

In 2009, Pope and coworkers described the synthesis of 2,4,6-trisubstituted pyrimidines by tandem oxidation/heterocyclocondensation of propargylic alcohols and amidines, which was effected rapidly and efficiently under microwave dielectric heating using barium manganate as oxidant, with microwave irradiation at 150°C in a mixture of EtOH-acetic acid for 45 min (Scheme 2.13) [15]. It was noticing that a tandem oxidation/heterocyclocondensation reaction offered a viable route to a range of *p*-extended pyrimidines. Furthermore, in all cases, the refined conditions using BaMnO$_4$ gave significant improvements in yield over the other methods making it by far the more efficient procedure. In order to broaden the electronic profile of the π-extended pyrimidines accessible by this methodology, bromides were transformed by copper-mediated N-arylation using the preformed Cu(I) catalyst Cu(neocup) (PPh$_3$)Br. This method can be further derivatized by microwave-assisted copper-mediated N-arylation. The π-extended pyrimidines so-formed by this approach were highly fluorescent in the visible region, displaying solvent-dependent emission wavelengths suggestive of charge transfer-dominated excited states.

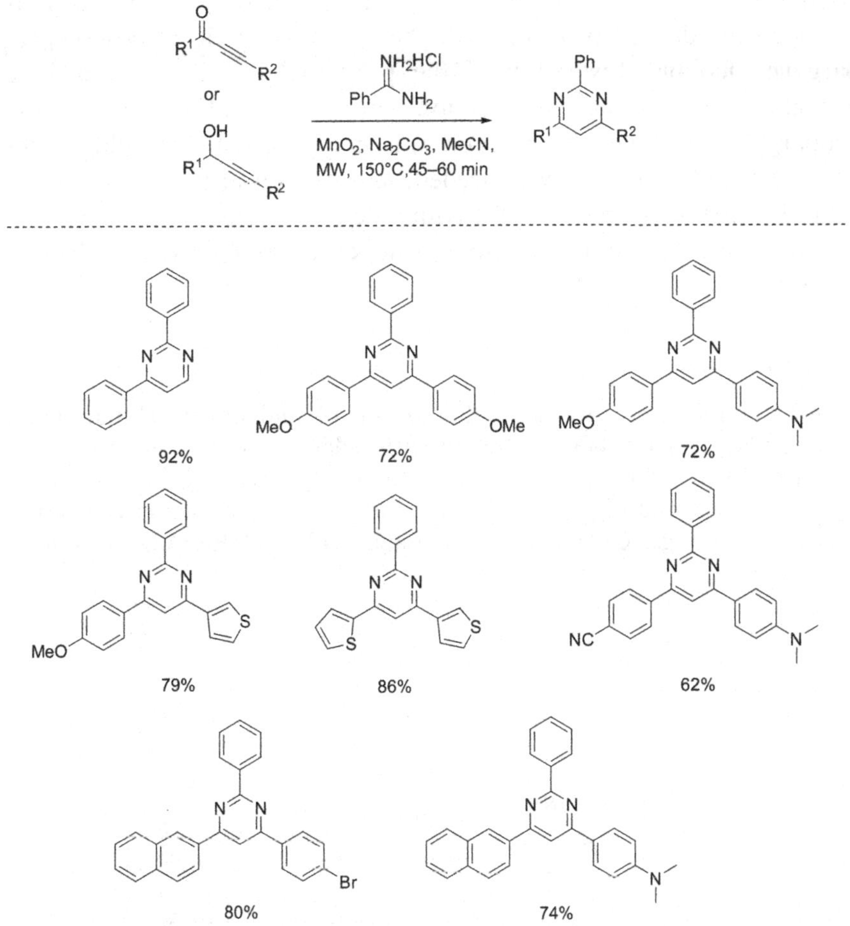

Scheme 2.13 Synthesis of pyrimidines using barium manganate.

In 2011, a new approach to the tandem synthesis of 2,4-disubstituted or 2,4,6-trisubstituted pyrimidines from propargylic alcohols with amidine was described by Zhan and coworkers (Scheme 2.14) [16]. In this article, they used $Cu(OTf)_2$ as the catalyst and gave the pyrimidines in moderate to good yields via a propargylation–cyclization–oxidation tandem sequence. This reaction can be carried out under mild conditions and oxidized by atmospheric air. Besides, water is the only by-product. A slower reaction rate and lower yield was observed when the catalytic amount of $Cu(OTf)_2$ was decreased to 10 mol%. Various propargylic alcohols, including secondary phenyl-substituted propargylic alcohols and terminal propargylic alcohols, underwent this tandem reaction to give the pyrimidines in moderate to good yields. They also found that electron-rich propargylic alcohols provided the desired products in higher yields than electron-poor propargylic alcohols. Instability of the propargylic cation intermediate probably made the tandem reaction less favorable. For the aromatic propargylic alcohols, the reaction was completed under mild conditions. No added oxidant was needed, and oxygen in the air was sufficient to ensure formation of pyrimidines. A majority of functional groups, such as bromo, methoxy, and cyclohexenyl, were tolerated under the reaction conditions.

Scheme 2.14

To a solution of propargylic alcohol (0.5 mmol) and amidine (1 mmol) in PhCl (2 mL), $Cu(OTf)_2$ (0.1 mmol) was added, and it was stirred at reflux. When the reaction was completed (monitored by TLC), the solvent was removed under vacuum, and then the residue was further purified by silica gel column chromatography (PE and EtOAc) to afford pyrimidine.

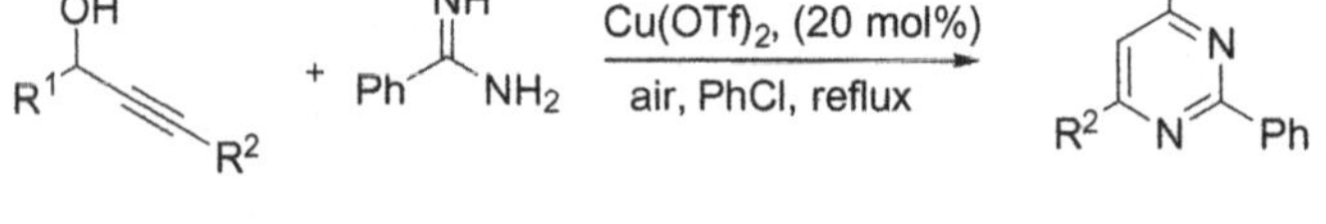

Scheme 2.14 Copper(II)-catalyzed synthesis of pyrimidines.

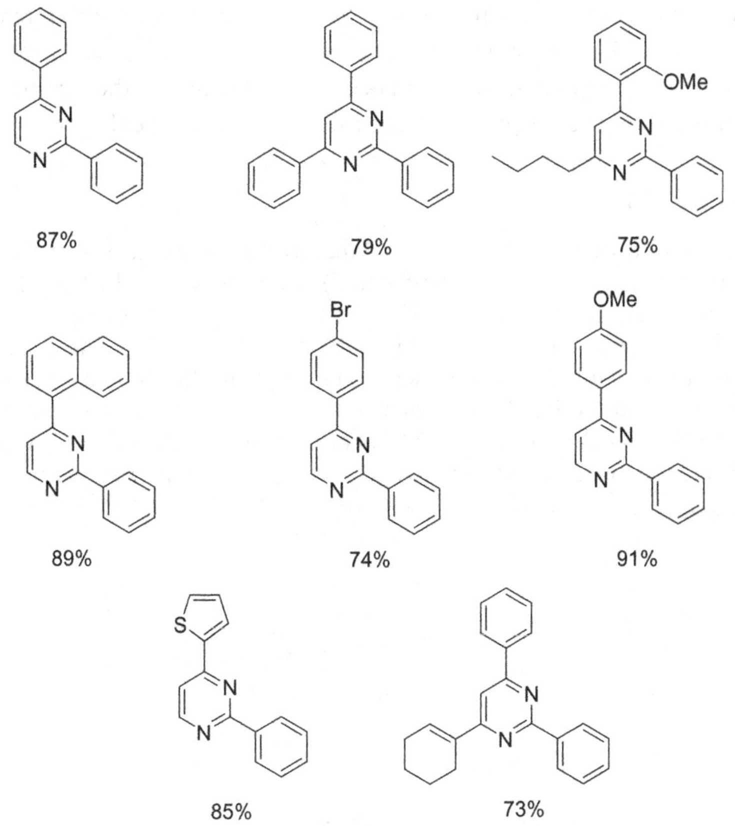

87%   79%   75%

89%   74%   91%

85%   73%

Scheme 2.14 (Continued)

In 2012, a novel cascade reaction for the synthesis of diverse members of pyrimidines was developed by Hu and coworkers (Scheme 2.15) [17]. They used the prepared 3-chlorochromenones and various amidines as the starting materials. The results showed that 3-chlorochromones with electron-withdrawing groups or electron-donating groups could react smoothly under the optimal conditions to form the pyrimidines in moderate to good yields. In contrast, changes in the R^2 amidine substituent were observed to have a profound impact on the efficiencies of the process. Specifically, reactions of amidines with the R^2 substituent being an alkyl group or variously substituted aryl groups occur in good to excellent yields. However, when R^2 is an electron-donating group, the yields of pyrimidines would be low. Interestingly, structurally and functionally complicated polyheterocyclic scaffolds could be constructed by using the new process. This tandem reaction was promoted by using a simple copper(I) reagent and involved a chemoselective Michael addition heterocyclization–intramolecular cyclization sequence. Significantly, the conditions utilized for the tandem process were mild and economical.

Scheme 2.15

To a solution of DBU (0.18 mL, 1.2 mmol) was added a mixture of 3-chlorochromone (100 mg, 0.55 mmol), acetamidine hydrochloride (57.6 mg, 0.61 mmol), CuBr (15.9 mg, 0.11 mmol), and 1,10-phenanthroline monohydrate (72.5 mg, 0.33 mmol) in DMF (3 mL). The mixture was heated to 90°C for 10 h under N_2 atmosphere, then cooled to room temperature, and diluted with water (40 mL). The mixture was extracted with DCM (Dichloromethane) (3 × 40 mL). The extracts were dried over Na_2SO_4 and concentrated in vacuo to give a residue that was purified by silica gel.

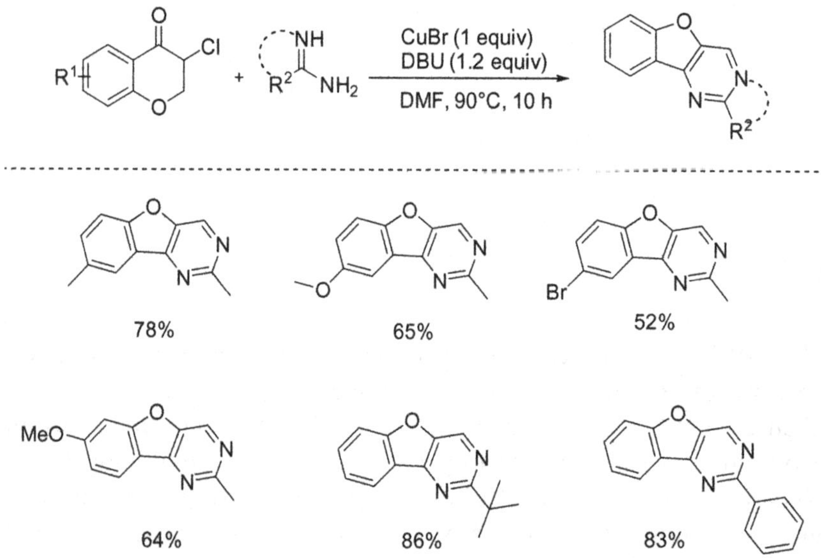

Scheme 2.15 Copper(I)-mediated cascade reactions for synthesis of pyrimidines.

In 2012, Batra and coworkers reported an efficient copper-catalyzed cascade reaction of 4-iodopyrazolecarbaldehydes and 4-iodopyrazolecarboxamides with substituted amidines for the preparation of substituted pyrimidines (Scheme 2.16) [18]. Most substrates examined afforded the required products in good yields. It was observed that changes in the substitution at the pyrazole ring did not have any deleterious effect on the outcome. Acetamidine gave the products with relatively better yields, whereas the (pyridine-4-yl)acetamidine generated the desired products with slightly lower yields. Nevertheless, from the study it may be inferred that the protocol was general and worked efficiently with a wide range of substrates.

Scheme 2.16

To a solution of 4-iodo-1,5-diphenyl-1*H*-pyrazole-3-carbaldehyde (250 mg, 0.67 mmol) and acetamidine hydrochloride (75 mg, 0.80 mmol) in DMSO (4 mL), Cs_2CO_3 (652 mg, 2.00 mmol), CuI (13 mg, 0.067 mmol), and l-proline (15 mg, 0.134 mmol) were added and the reaction mixture was heated at 90°C for 12 h under a nitrogen atmosphere. Thereafter, water (50 mL) and ethyl acetate (25 mL) were added and the reaction mass was pass through a celite bed and the layers were separated. The aqueous layer was further extracted with ethyl acetate (2 × 20 mL) and the collected organic layer was washed with brine, dried with anhydrous Na_2SO_4, concentrated under vacuum, and purified by silica gel.

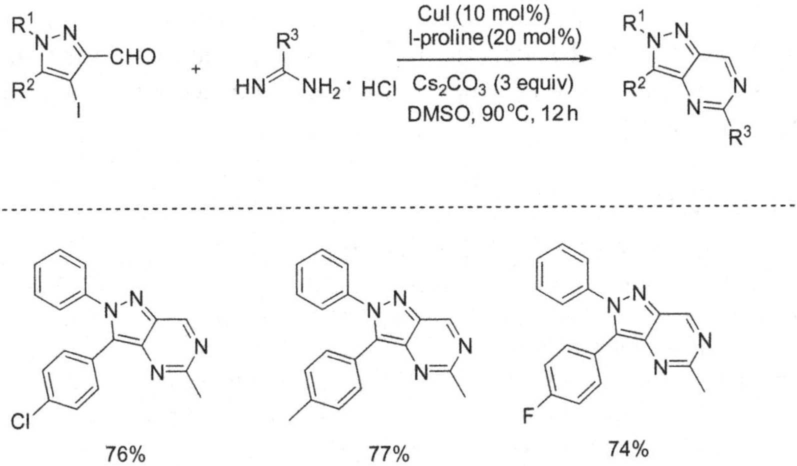

76%   77%   74%

Scheme 2.16 Copper-catalyzed synthesis of pyrimidines.

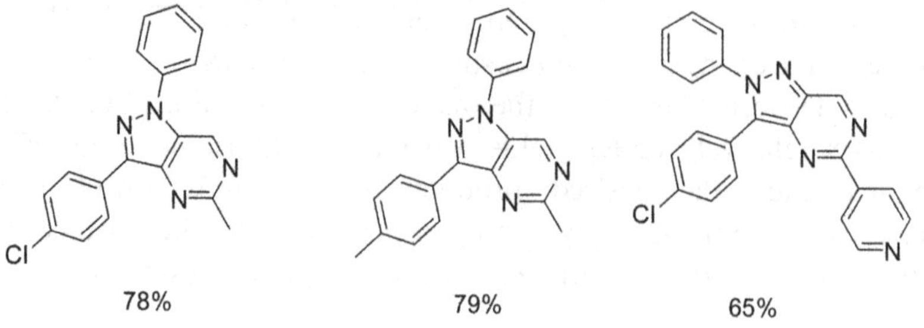

78%   79%   65%

Scheme 2.16 (Continued)

In 2014, Xie and coworkers developed a novel methodology for the synthesis of fully substituted pyrimidines from commercially available amino acid esters with Cu(II)/TEMPO as the promotor (Scheme 2.17) [19]. In this reaction, the amino acid esters acted as the only N–C sources for the construction of corresponding pyrimidines. The mechanism of this process includes oxidative dehydrogenation, the generation of an imine radical, and a formal [3 + 3] cycloaddition. This methodology proved to be a high atom-economic and straightforward strategy for the synthesis of pyrimidines. And many substrates which were substituted by various functional groups were afforded in moderate to good yield. Notably, for substrates bearing electron-withdrawing or weakly electron-donating *para*-substituents on the benzene ring. The desired pyrimidines were obtained in good yield with methyl, ethyl, or benzyl substituents. Besides, the reaction was influenced by the steric effects. The process was simple, efficient, atom-economic, and had great practical worth.

Scheme 2.17

To a solution of the amino acid ester (0.28 mmol) in anhydrous xylene (10 mL) was added the $Cu(OAc)_2$ (0.28 mmol, 1.0 equiv), TEMPO (0.56 mmol, 2.0 equiv), and NH_3-SiO_2 (100% mass fraction of amino acid ester) under argon. After stirring in refluxing xylene for 2.0 h, the mixture was cooled to room temperature and the solvent removed by evaporation under reduced pressure. The resulting residue was purified by flash column chromatography to give the pure pyrimidine product.

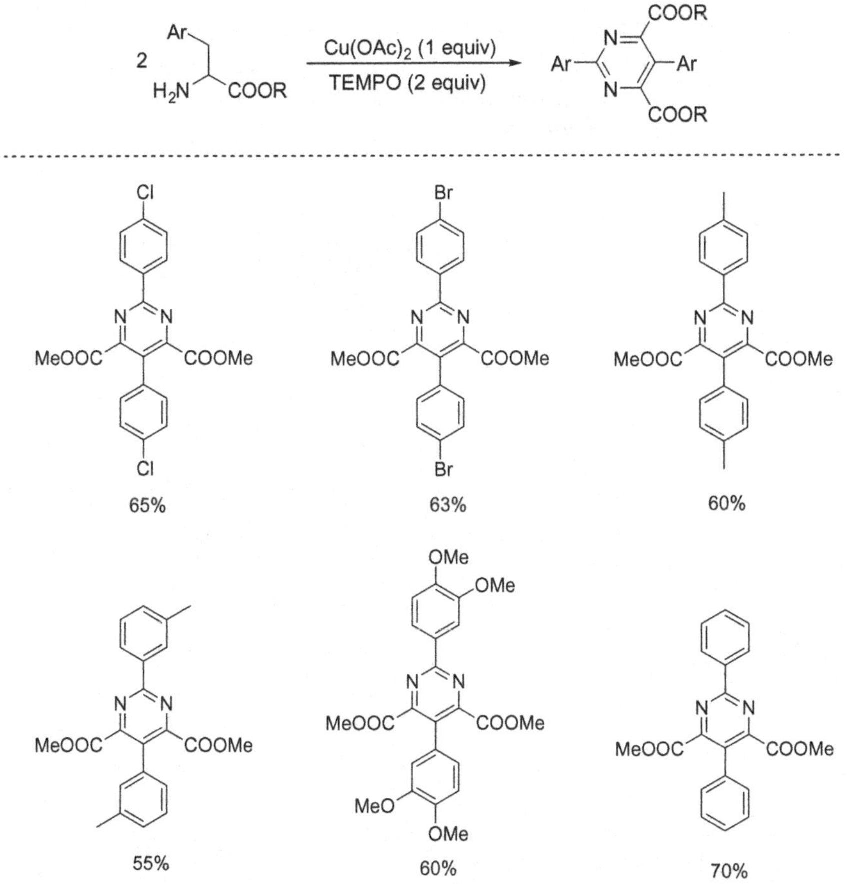

Scheme 2.17 *Cu(II)/TEMPO-promoted one-pot synthesis of pyrimidines.*

In 2016, Cao and coworkers provided a general and efficient Au-catalyzed domino intramolecular cyclization process for the synthesis of 2,4-disubstituted pyrimidines in CH_2Cl_2 at the temperature (Scheme 2.18) [20]. This transformation gave a new method for the formation of C−C and C−N bonds via intramolecular cyclization. And this method provided a simple route to prepare pyrimidines which were broadly applicable for the synthesis of biologically active molecules.

Scheme 2.18

A reaction tube was charged with 3-Phenylpropiolaldehyde (0.5 mmol), cyclopropanecarboximidamide (0.6 mmol), Ph_3PAuCl (3 mol%), K_2CO_3

(1.0 mmol), and CH_2Cl_2 (3 mL). The mixture was stirred for 3 h at room temperature. After reaction completion, as monitored by TLC and GC-MS analysis, the solvent was removed and the crude product was separated by column chromatography.

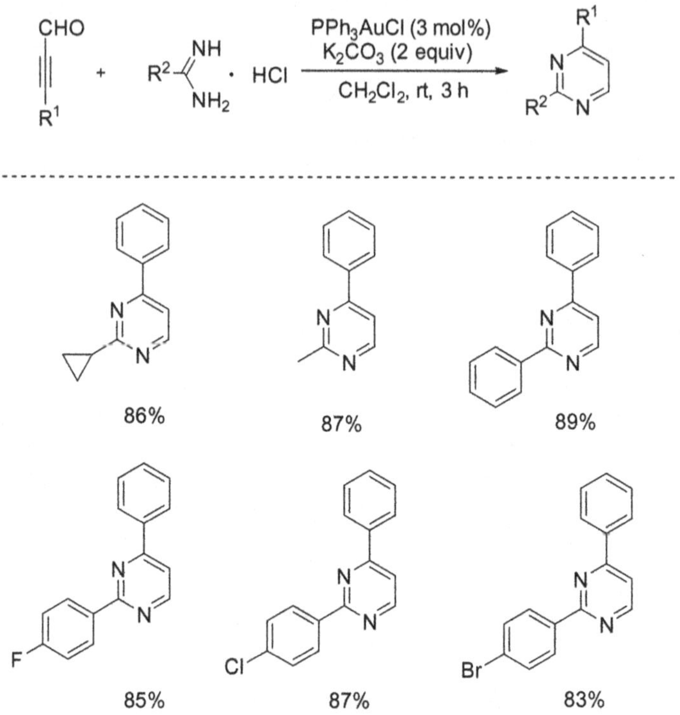

Scheme 2.18 Au-catalyzed synthesis of pyrimidines.

2.4 OTHER PROCEDURES FOR PYRIMIDINE SYNTHESIS

In 2000, Zieliński and coworkers reported that a new group of 6- and 7-substituted compounds of 4-amino-2-phenylquinazoline can be synthesized by the reaction of N-arylbenzimidoyl chlorides with cyanamide in the presence of $TiCl_4$ (Scheme 2.19) [21]. First, N-arylbenzamides were reacted with PCl_5 in benzene at elevated temperatures to afford the corresponding N-arylbenzimidoyl chlorides. Then, N-arylbenzimidoyl chlorides was treated with cyanamide at room temperature, yielding intermediates 3. Finally, after several hours of heating in benzene in the presence of the Lewis acid catalyst $TiCl_4$, 3 underwent cyclization to give the final product.

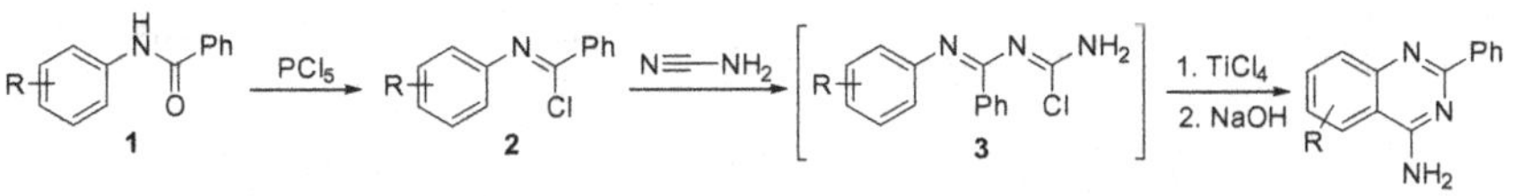

Scheme 2.19 Synthesis of pyrimidine by three steps.

In 2005, a mild, practical procedure for oxidative dehydrogenation with catalytic amounts of a Cu salt, K_2CO_3, and tert-butylhydroperoxide (TBHP) as a terminal oxidant to synthesis pyrimidines was developed by Yamamoto and coworkers (Scheme 2.20) [22]. This oxidation procedure was generally applicable to dihydropyrimidinones and most dihydropyrimidines. These conditions are applicable to alkyl and aryl substituents with a range of electronic properties. Notably, oxidatively labile functionalities such as thioether and amines were not affected.

Scheme 2.20

A reactor was purged with nitrogen and charged with dihydropyrimidine (700 g, 2.39 mol, 1.0 equiv), $CuCl_2$ (3.25 g, 0.024 mol, 1.0 mol%), K_2CO_3 (33.1 g, 0.243 mol, 0.1 equiv), and CH_2Cl_2 (7.0 L). The suspension was heated to 35°C and treated with tert-butylhydroperoxide (70% aqueous solution) (635.0 g, 677.2 mL, 2.0 equiv) over 120 min with vigorous agitation.

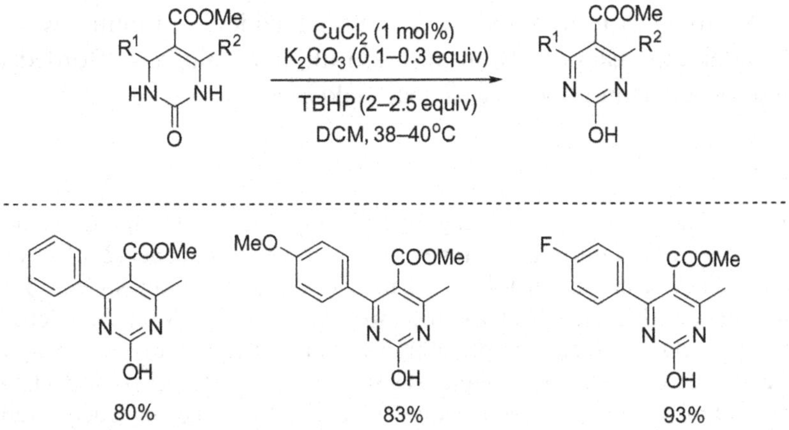

Scheme 2.20 Synthesis of pyrimidine.

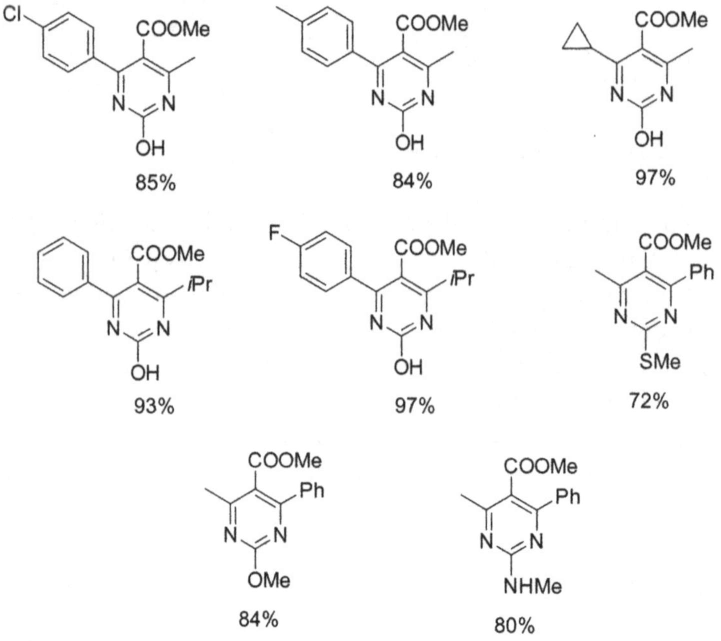

Scheme 2.20 (Continued)

In 2009, a one-pot synthesis of *N*-sulfonyl-2-alkylidene-1,2,3,4-tetra-hydropyrimidines via a highly selective and copper-catalyzed multi-component reaction of sulfonyl azides, terminal alkynes, and α,β-unsaturated imines was developed by Wang and coworkers (Scheme 2.21) [23]. The α,β-unsaturated imine substrates could be generated from amines and a,b-unsaturated aldehydes in a one-pot process. A variety of terminal alkynes, sulfonyl azides, and imines worked well under this reaction. Both aryl and aliphatic alkynes afforded the desired products in good to excellent yields.

Scheme 2.21

To a mixture of CuBr (14 mg, 0.1 mmol), alkynes (1 mmol), azides (1 mmol), and unsaturated imines (1.05 mmol) in toluene (2 mL) was added TEA (2 mmol) in toluene (1 mL) under an N_2 atmosphere. The mixture was then stirred at room temperature for 3 h. After completion of the reaction, which was monitored by TLC, the mixture was evaporated under vacuum. The residue was subjected to silica gel column chromatography with petroleum ether/ethyl acetate. The products were recrystallized from petroleum ether/ethyl acetate.

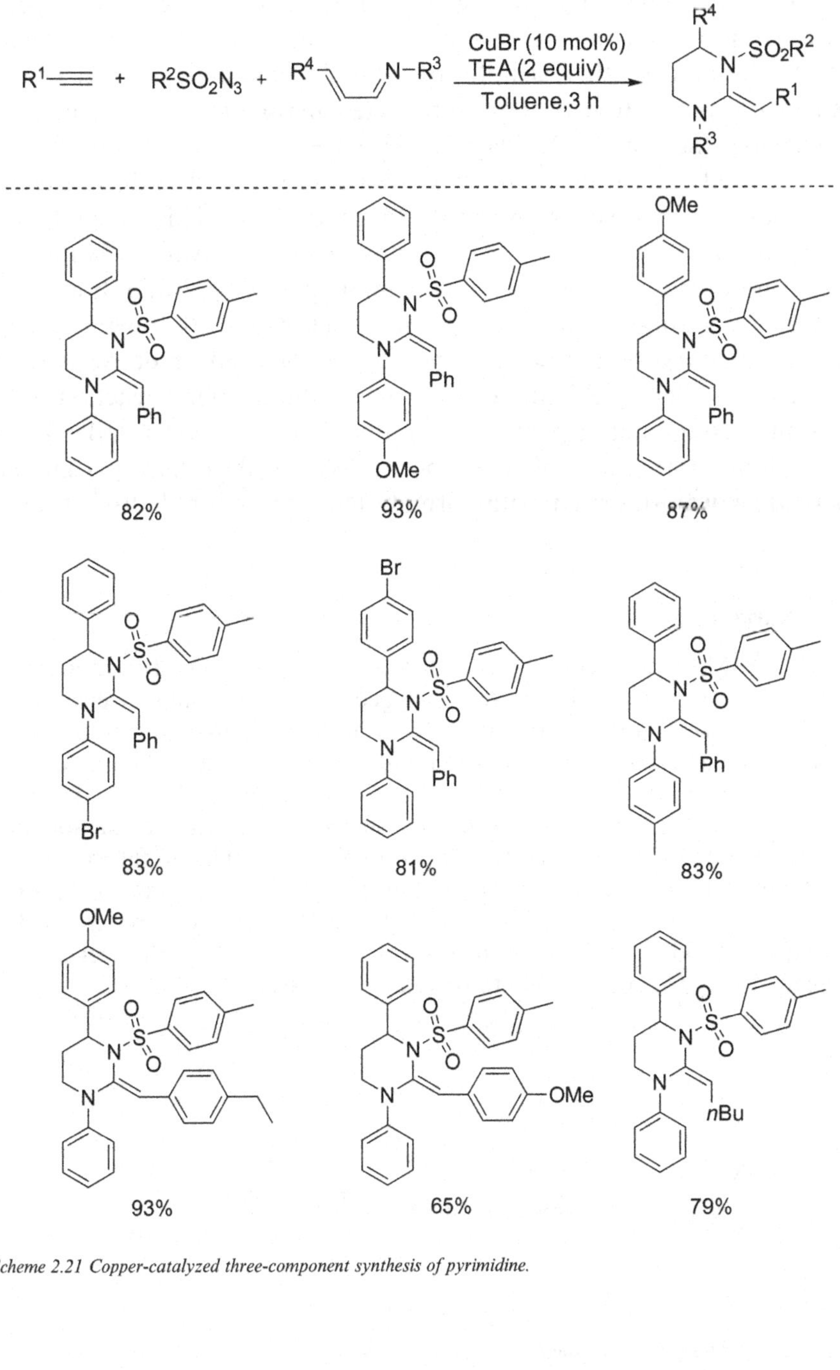

Scheme 2.21 *Copper-catalyzed three-component synthesis of pyrimidine.*

In 2009, Konakahara and coworkers reported a ZnCl$_2$-catalyzed three-component coupling reaction involving a variety of functionalized enamines, triethyl orthoformate, and ammonium acetate, which led to the production of 4,5-disubstituted pyrimidine derivatives in a single step (Scheme 2.22). [24] This [3 + 1 + 1] annulation process worked well and produced the corresponding mono- and disubstituted pyrimidine derivatives in moderate to good yields. This method successfully accommodated other acetophenone derivatives with an electron-donating group and an electron-withdrawing group. Enamines with an electron-donating group and an electron-withdrawing group on the benzene ring produced the desired pyrimidine derivatives in good to excellent yield. In contrast, when the coupling reaction with enamine, containing a pyridin-2-yl group, was performed under these conditions, formation of the unexpected trisubstituted pyrimidine derivative was observed with formation of the desired disubstituted pyrimidine.

Scheme 2.22

To a PhMe (1 mL) solution of ZnCl$_2$ (6.8 mg, 0.050 mmol) and acetal (220 mg, 1.5 mmol) in a screw-capped vial were added enamine (0.50 mmol) (or ketone) and ammonium acetate (77 mg, 1.0 mmol), and the vial was sealed with a cap containing a PTFE (polytetrafluoroethene) septum. The mixture was heated at 100°C and monitored by TLC or GC analysis until the enamine had been consumed. To quench the reaction, a saturated aqueous solution of NaHCO$_3$ (5 mL) was added to the mixture. The mixture was extracted several times with CHCl$_3$, and the combined organic extracts were dried over Na$_2$SO$_4$, filtered, and then concentrated under reduced pressure. The crude product was purified by silica gel chromatography (AcOEt-hexane) to produce pyrimidine.

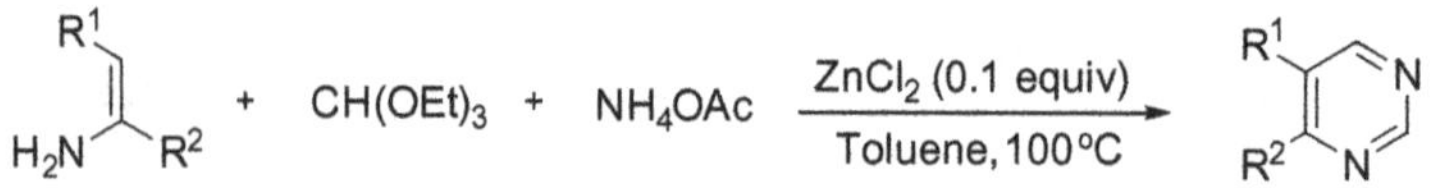

Scheme 2.22 ZnCl$_2$-catalyzed three-component synthesis of pyrimidine.

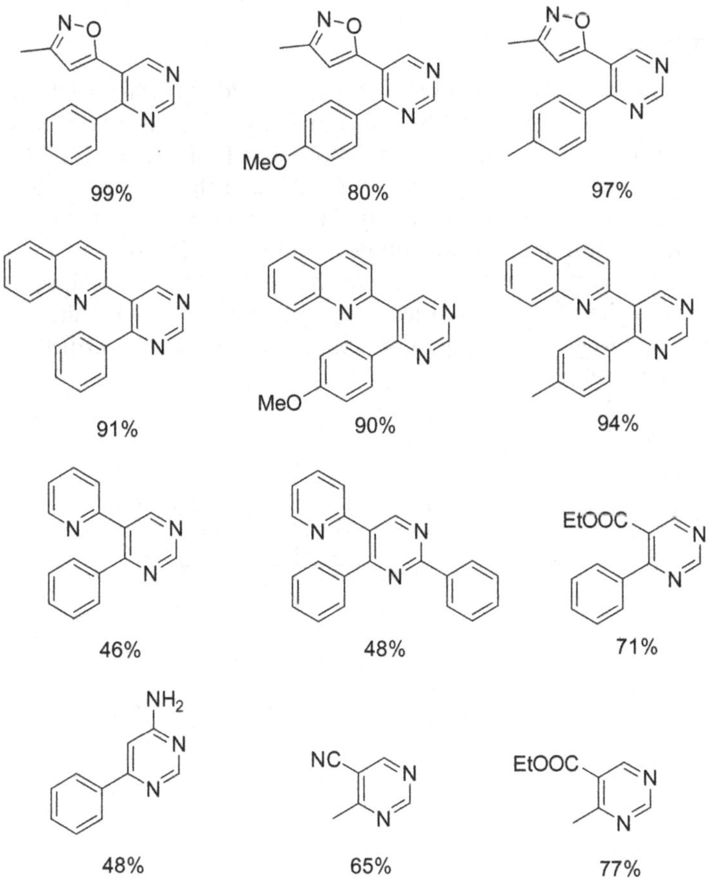

99%   80%   97%

91%   90%   94%

46%   48%   71%

48%   65%   77%

Scheme 2.22 (Continued)

In 2010, Hu and coworkers developed a three-component one-pot approach to synthesize pyrimidines, which was through from iodochromone, alkyne, and an amidine through a Sonogashira coupling, condensation, and cycloaddition (Scheme 2.23) [25]. In this reaction, on changing the electronic and steric properties (R^2) on the acetylene moiety the corresponding products were afforded in moderate to good yields. An electron-donating group (R^1) at the 6-position or 7-position of iodochromone gave the corresponding product in a reasonable yield. Apparently, an electron-withdrawing group such as NO_2 or Br at the 6-position of iodochromone afforded complicated products in very low yields. It was worth noting that amidines with an electron-donating group preceded the formation of the desired product in one-pot tandem process smoothly.

Scheme 2.23

Iodochromone (0.2 mmol), alkyne (1.5 equiv), $PdCl_2(PPh_3)_2$ (0.01 mmol), CuI (0.02 mmol), amidine (1.5 equiv), and mixed bases of DIPEA (2.0 equiv) and K_2CO_3 (4.0 equiv) were dissolved in DMF (2.0 mL). The mixture was stirred at room temperature for 2 h and then heated at 60°C for 6 h. The reaction was monitored by TLC. After the reaction was complete, the resulting mixture was diluted with water (20 mL) and extracted with ethyl acetate (25 mL × 3), and the combined organic layers were washed with brine (20 mL), dried over anhydrous Na_2SO_4, filtered, and concentrated to give the crude product, which was further purified by column chromatography.

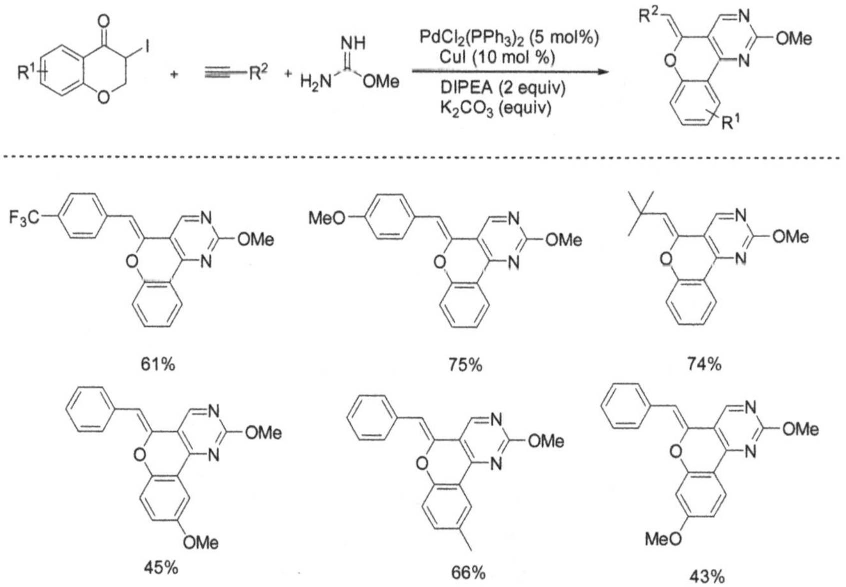

Scheme 2.23 Three-component one-pot approach to synthesize pyrimidines.

In 2010, titanium catalyzed one-pot multicomponent coupling reactions for direct access to substituted pyrimidines were presented by Odom and coworkers (Scheme 2.24) [26]. In this reaction, they provided 17 examples of pyrimidines using this one-pot, four-component procedure from simple starting materials. And in some cases, catalyst architecture could be tuned to control the regioselectivity of the alkyne addition. The multicomponent coupling reaction was quite effective for both terminal and internal alkynes with a variety of aliphatic and aromatic amines. Moreover, heteroaromatic alkynes as well as enynes could be successfully converted to the corresponding pyrimidine compounds.

Scheme 2.24

In a N_2 filled glove box, a 40 mL pressure tube, equipped with a magnetic stirbar was loaded with amine (5 mmol), catalyst (10−20 mol%), alkyne (5 mmol), isonitrile (5−7.5 mmol), and 10 mL of dry toluene. The pressure tube was sealed with a Teflon screw cap, taken out of the dry box, and heated to the appropriate temperature for the desired time with vigorous stirring. After completion of the reaction as judged by GC-FID, the pressure tube was cooled to room temperature and volatiles were removed under reduced pressure. Then the same pressure tube was charged with amidine hydrochloride (7.5 mmol) in tert-amyl alcohol (10 mL) and heated to 150°C for 24 h. After completion of the reaction, tert-amyl alcohol was removed under reduced pressure. The crude product was dissolved in CH_2Cl_2 and washed with water. The organic layer was dried over Na_2SO_4 and concentrated on a rotary evaporator. The crude product was purified either by column chromatography or by crystallization from a suitable solvent.

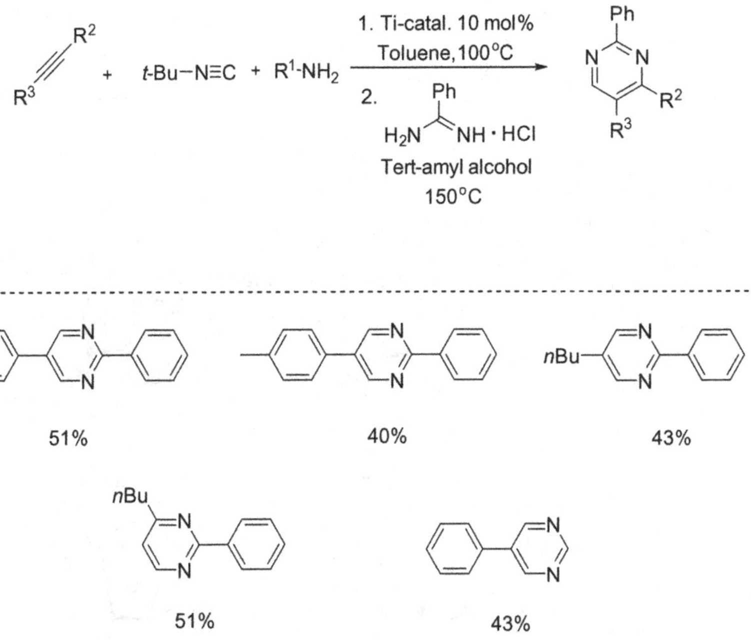

Scheme 2.24 Titanium catalyzed one-pot reactions to synthesize pyrimidines.

In 2011, Yu and coworkers reported an efficient gold(I)-catalyzed approach to the synthesis of pyrimidines (Scheme 2.25) [27]. In this reaction, readily accessible and shelf-stable glycosyl *ortho*-hexynylbenzoates were shown to be superior under the catalysis of [Ph₃PAuNTf₂]. The success of this highly efficient and regioselective N'-glycosylation of purines could be attributed to the mild glycosylation conditions that enable Boc-protected purine derivatives to be used as coupling partners.

Scheme 2.25

To a stirred suspension of 2,3,5-tri-O-benzoyl-β-D-ribofuranosyl *ortho*-hexynylbenzoate (1.29 g, 2.0 mmol), uracil (269 mg, 2.4 mmol) in dry CH_3CN (2 mL) was added BSTFA (2.62 mL, 9.6 mmol) under argon atmosphere. The mixture was stirred at room temperature for 30 min; a clear solution resulted. $Ph_3PAuNTf_2$ (148 mg, 0.2 mmol) was added, the stirring was continued at room temperature for 3 days. The solvent was removed under reduced pressure, the resulting residue was purified by silica gel.

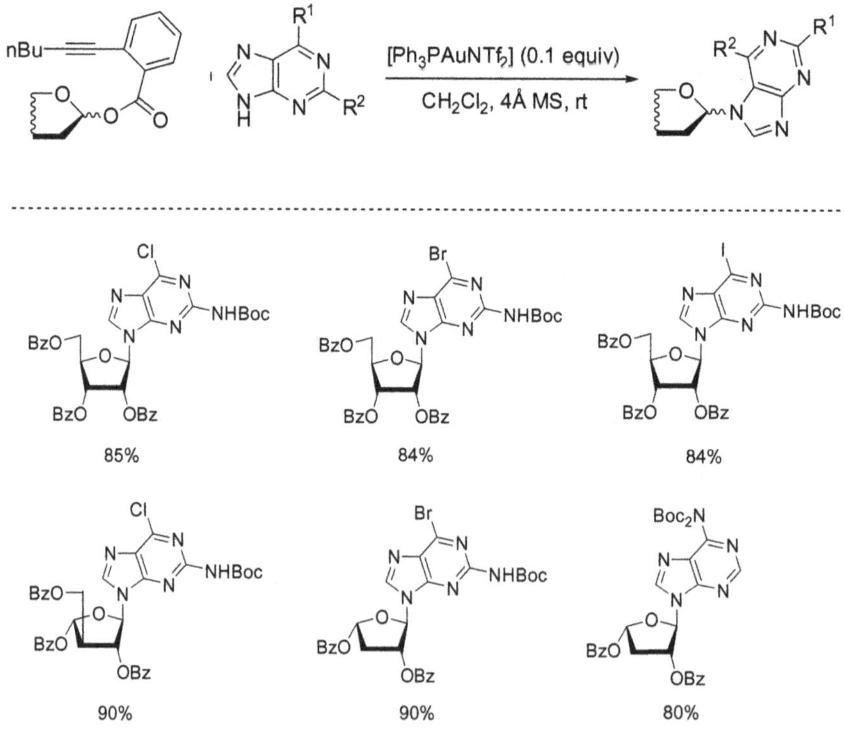

Scheme 2.25 Gold(I)-catalyzed approach to the synthesis of pyrimidines.

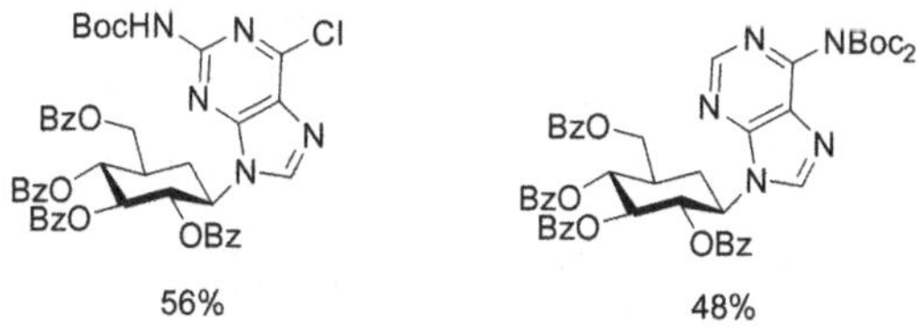

56%   48%

Scheme 2.25 (Continued)

In 2011, Han and coworkers reported dihydropyrimidines were oxidized to the corresponding pyrimidines, which were in high yields by molecular oxygen in the presence of catalytic amount of *N*-hydroxyphthalimide (NHPI) and Co(OAc)$_2$ in a mild and environmental benign condition (Scheme 2.26) [28]. Substrates with electron-withdrawing or electron-donating groups gave excellent yields of the products under this conditions. However, substrates with *t*Bu, *i*Pr, or Ph gave very poor yields.

Scheme 2.26

A mixture of dihydropyrimidine (290 mg, 1 mmol), NHPI (16 mg, 10 mol%), and Co(OAc)$_2$ 4H$_2$O (1 mg, 0.5 mol%) was placed in a 25 mL three-necked flask in (CH$_2$Cl)$_2$ (4 mL) and stirred at 80°C under oxygen atmosphere for 1 h. When the starting materials were consumed completely monitored by TLC, the reaction mixture was concentrated by vacuum and then the product was isolated by silica gel column chromatography.

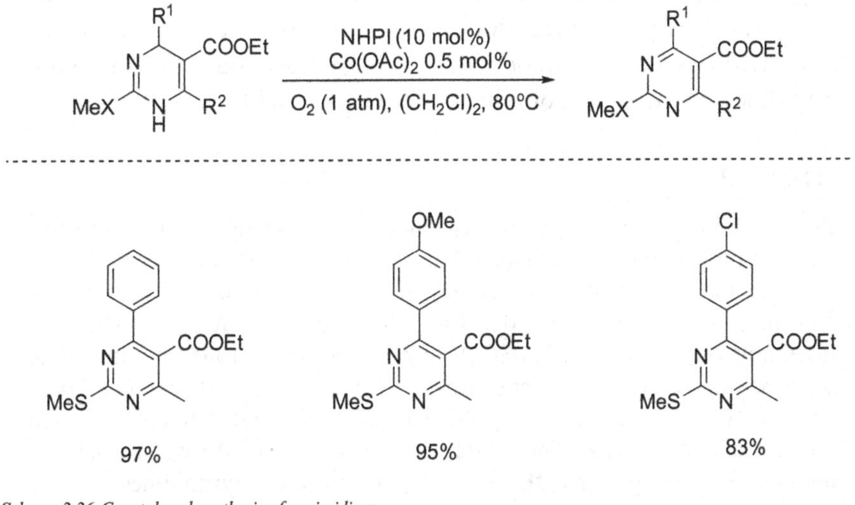

97%   95%   83%

Scheme 2.26 Cocatalyzed synthesis of pyrimidines.

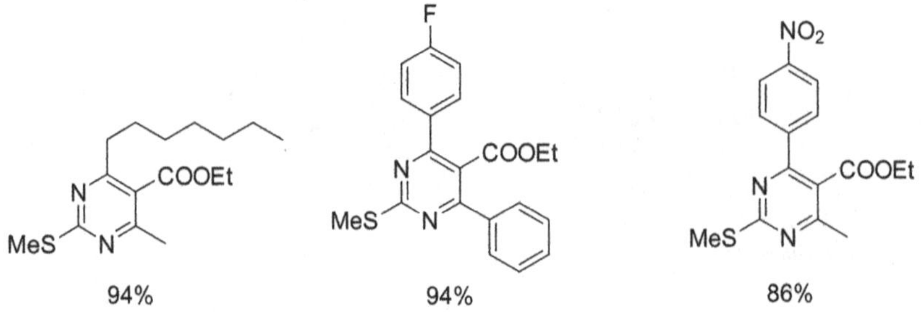

Scheme 2.26 Cocatalyzed synthesis of pyrimidines.

In 2011, Konakahara and coworkers had developed procedures for the synthesis of tri- and tetrasubstituted pyrimidine derivatives by an unprecedented [5 + 1] annulation using enamidines, which was under solvent-free reaction conditions (Scheme 2.27) [29]. This reaction led to the synthesis of a variety of polysubstituted pyrimidine derivatives in good yields. The formation of multisubstituted pyrimidines was accomplished by the $ZnBr_2$-catalyzed annulation of the enamidine with an orthoester as the C1 unit. Moreover, they had demonstrated the skeletal transformation of the pyrimidine skeletons containing an isoxazolyl group on the C5 atom into pyrido [2,3-d]pyrimidin-5-one frameworks through the reductive ring opening of the isoxazole ring with $[Mo(CO)_6]$, followed by intramolecular cyclization with t-BuOK in excellent yield. When enamidines containing an electron-donating or electron-withdrawing group on the benzene ring were employed, the corresponding pyrimidine derivatives were obtained in good to excellent yield. Unfortunately, when the reaction of enamidine with tetraethoxymethane was carried out under our previously optimized conditions, the desired pyrimidine derivative was obtained in only 18% yield.

Scheme 2.27

$ZnBr_2$ (4.5 mg, 0.020 mmol) was added to a solution of enamidine (0.200 mmol) and orthoester (0.400 mmol) in PhMe (0.4 mL) in a screw-capped vial, and the vial was sealed with a cap containing a PTFE septum. The mixture was heated at 110°C for 72 h. To quench the reaction, a saturated aqueous solution of $NaHCO_3$ (5 mL) was added to the mixture. The mixture was extracted several times with $CHCl_3$, and the combined organic extracts were dried over Na_2CO_3, filtered, and then concentrated under reduced pressure. The crude product was purified by silica gel column chromatography (AcOEt/hexane) to produce the pyrimidines.

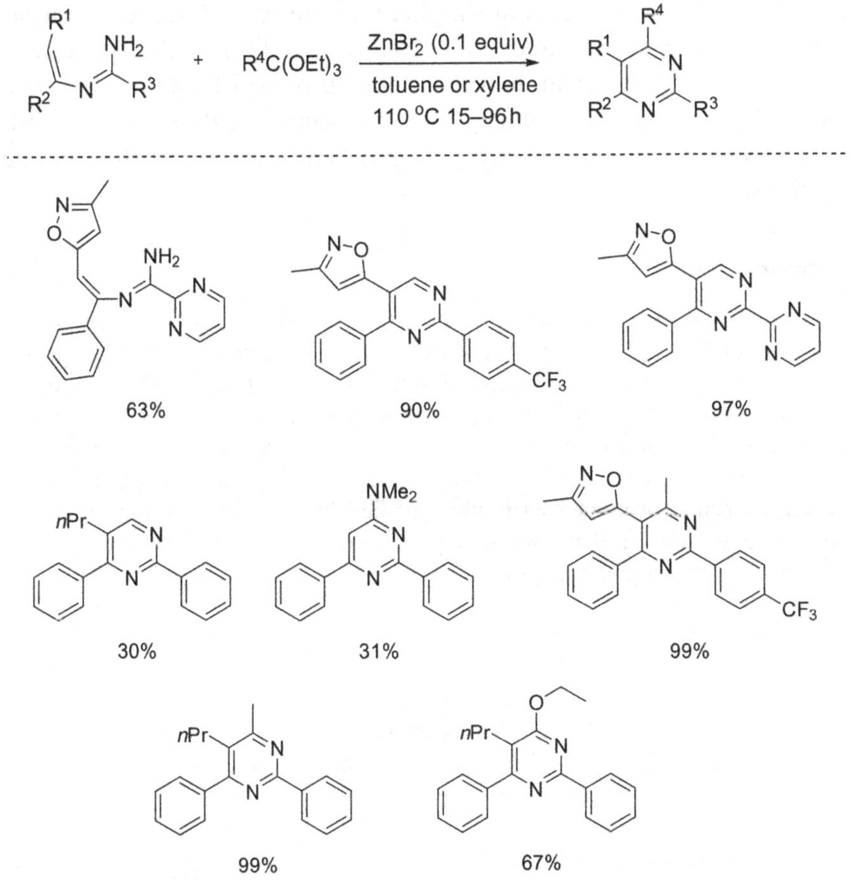

Scheme 2.27 Synthesis of tri- and tetrasubstituted pyrimidines.

In 2012, a novel synthetic protocol for the one-pot chemo- and stereoselective construction of diversely functionalized pyrimidines via copper(I)-catalyzed cycloaddition of sulfonyl azides, alkynes and N-arylidenepyridin-2-amines under mild reaction conditions was reported by Pitchumani and coworkers(Scheme 2.28) [30]. In addition, the catalytic activity of copper(I)-modified zeolite, a recyclable, heterogeneous catalyst was also investigated in this reaction, which gave improved yield compared to its homogeneous equivalents. The attractive features of this protocol were the ready aromatization without any oxidizing agents, the reversal of chemoselectivity, the excellent stereoselectivity, and a new method for the synthesis of an important pharmacophore, pyrido[1,2-a]pyrimidine-4-imine, with a diverse range of biological activities and applications at room temperature. This methodology

offered advantages in terms of simplicity of the procedure, ready varia-
tion in building blocks, and mild reaction conditions. Both electron-
rich and electron-deficient sulfonyl azides provided higher yields. The
imines derived from aryl aldehydes with electron-withdrawing as well
as electron-donating groups are generally well tolerated in the cascade
reactions.

Scheme 2.28

A solution of alkyne (1 mmol) in one mL of DCM was added slowly to a
mixture of Cu(I)-zeolite (20 mg), sulfonyl azide (1 mmol), N-arylidenepyr-
idin-2-amine (1 mmol), and Et$_3$N (2 mmol) taken in 1 mL of DCM under
an air atmosphere. After stirring at room temperature for 12 h, the mixture was diluted with ethyl acetate (5 mL). After removing the catalyst by
filtration, followed by solvent evaporation under reduced pressure, the
resulting crude product was finally purified by column chromatography
on silica gel (60–120 mesh) with petroleum ether and ethyl acetate as
eluting solvent to give the desired product.

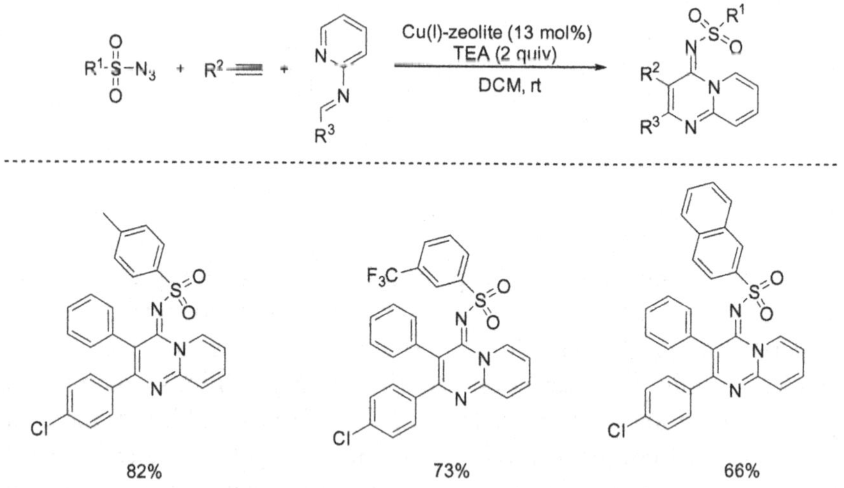

Scheme 2.28 Copper(1)-catalyzed one-pot synthesis of pyrimidines.

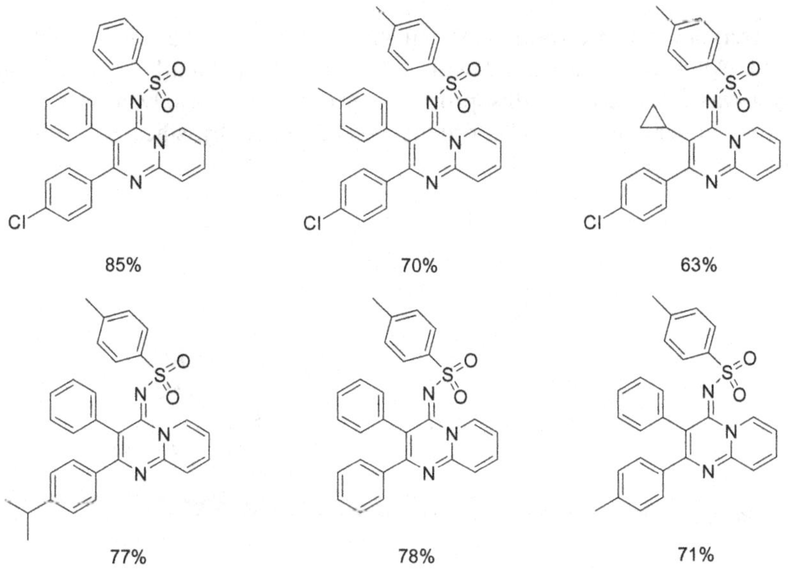

85%   70%   63%

77%   78%   71%

Scheme 2.28 (Continued)

In 2012, Obora and coworkers established a practical method for the preparation of polysubstituted pyrimidine derivatives, which gave moderate to good yields (Scheme 2.29) [31]. The regioselectivities of the desired products were influenced by electronic effects on the unsymmetrical internal alkynes. The reaction was successful using terminal alkynes with n-octyl, phenyl, and cyclohexyl groups. The terminal alkynes 1-decyne, phenylacetylene, and cyclohexylacetylene gave the corresponding desired products in 50–74% yields with >95% regioselectivity and excellent chemoselectivity. The reaction of an aliphatic nitrile such as octanonitrile or trimethylsilyl cyanide proved to be sluggish under these conditions. We proposed the reaction mechanism shown in Scheme 2.30. The reaction initiates $NbCl_5$-assisted reaction of nitrile to form N-benzylidenebenzamidine. Subsequent cycloaddition with phenylacetylene on the benzylidene carbon results in the formation of 2,4,6-triphenylpyrimidine.

Scheme 2.29

To a mixture of 4-octyne (110 mg, 1 mmol) and, benzonitrile (2 mL), $NbCl_5$ (1.2 mmol) was added in six batches (each 0.2 mmol) every 2 h, for 22 h at 60°C under Ar. The yields of products were estimated from the peak areas based on the internal standard technique using GC in 86% yield. After being quenched with 10% NaOH aq (50 mL), the organic layer was extracted with diisopropyl ether (30 mL). The solvent was evaporated under vacuum. The product was isolated by silica gel column chromatography.

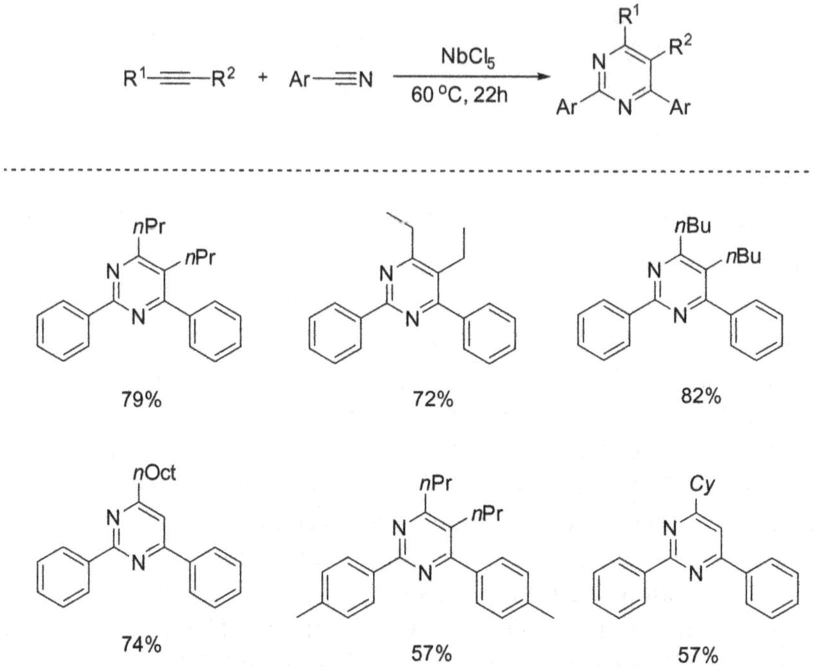
NbCl5
60 °C, 22h
79%
72%
82%
74%
57%
57%

Scheme 2.29 Nb-catalyzed synthesis of pyrimidines.

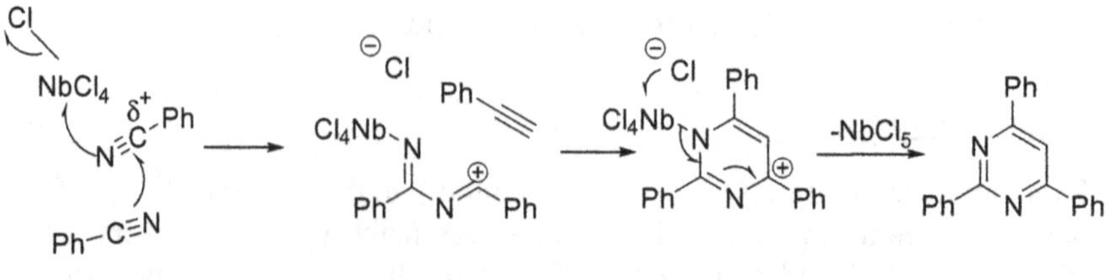
-NbCl5

Scheme 2.30 Mechanism for Nb-catalyzed synthesis of pyrimidines.

In 2013, Louie and coworkers reported a novel Fe-catalyzed cyclo-addition reaction between alkynenitriles and cyanamides to provide 2-aminopyrimidines (Scheme 2.31) [32]. In this reaction, catalytic amounts of FeI_2, $^{iPr}PDAI$, and Zn were found to effectively catalyze the [2 + 2 + 2] cycloaddition of a variety of cyanamides and alkyneni-triles to afford bicyclic 2-aminopyrimidines. What was especially remarkable about the reaction was that iron, which had traditionally been an inefficient cycloaddition catalyst for nitrile incorporation, could incorporate multiple nitriles into aromatic products. Various combinations of substrates gave good results. However, cyclic cyana-mides did not work well under this condition.

Scheme 2.31

In a nitrogen filled glove box, 5 mol% FeI_2 beads were added to the reac-tion vial. The beads were thoroughly crushed into a fine powder to ensure complexation with 10 mol% $^{iPr}PDAI$ and toluene was added to a vial and the mixture was stirred for 1 h. At this time additional toluene, a tol-uene solution of alkynenitrile, 30 mol% Zn dust and 3 equiv of cyana-mide were added. The vial was then capped, removed from the glove box, and stirred in a 40°C oil bath. After completion, as monitored by GC, the crude mixture was purified by silica gel flash chromatography.

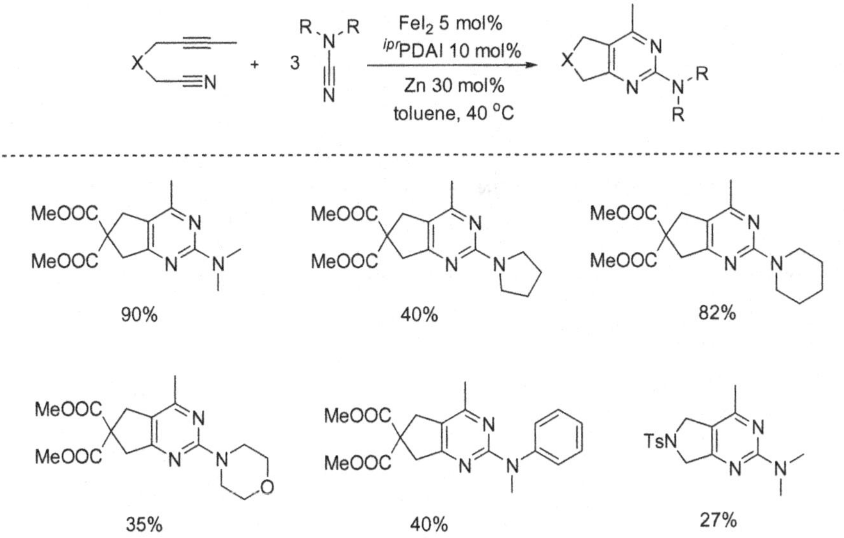

Scheme 2.31 Fe-catalyzed cycloaddition reaction.

In 2013, Liu and coworkers developed a new cycloaddition reaction of zirconocene butadiyne complexes with two molecules of aryl nitriles, which provided a rapid access to polysubstituted pyrimidines in a regioselective manner and in one pot (Scheme 2.32) [33]. A variety of aryl nitriles could be used as effective substrates for this reaction. It was found that the aromatic rings of nitriles bearing an electron-donating (Me, OMe) or an electron-withdrawing group (Cl, F, CF_3) were all compatible under the reaction conditions. The electronic nature of the aryl substituents on nitriles had little influence on the product yields. However, steric interactions affected the reaction.

Scheme 2.32

To a suspension of Cp_2ZrCl_2 (0.19 g, 0.65 mmol) in toluene (5 mL) was added dropwise n-BuLi (1.3 mmol, 0.81 mL, 1.6 M solution in hexane) at $-78°C$. After stirring for 1 h at the same temperature, 1,3-butadiyne (0.5 mmol) was added and the reaction mixture was warmed up to room temperature and stirred for 3 h. Aryl nitrile (1.5 mmol) was then added and the reaction mixture was gradually warmed up to 80°C. After the reaction was complete, as monitored by TLC, the resulting mixture was cooled down to room temperature, and filtered over celite. The filtrate was concentrated to give a viscous residue which was purified by the preparative thin layer chromatography to afford the desired products.

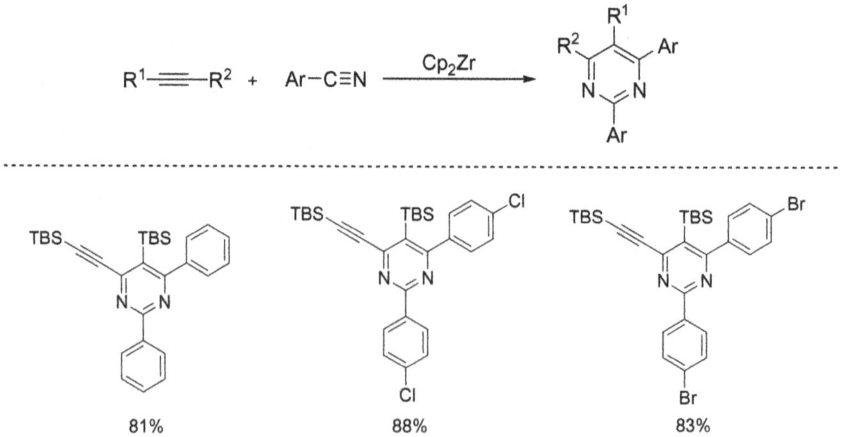

Scheme 2.32 Zr-catalyzed synthesis of pyrimidines.

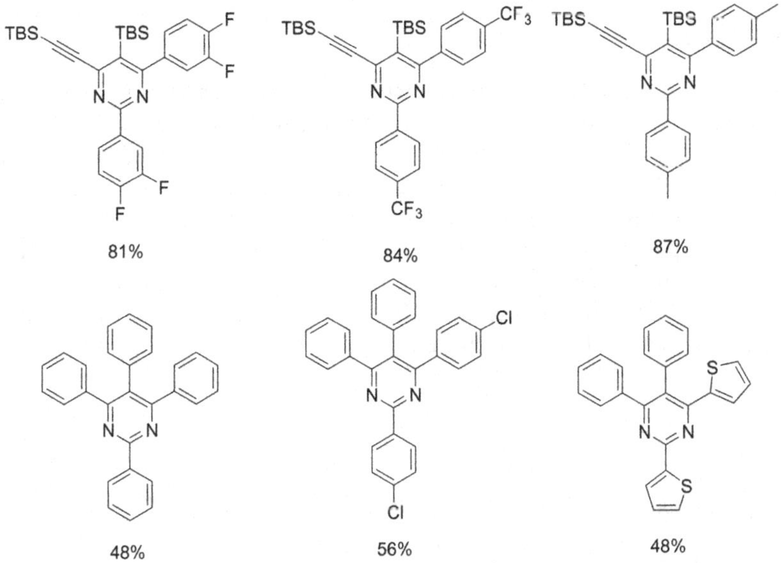

Scheme 2.32 Zr-catalyzed synthesis of pyrimidines.

In 2014, Liu and coworkers reported gold-catalyzed [2 + 2 + 2] cycloadditions between ynamides and nitriles to afford monomeric 4-aminopyrimidines, which were commonly found in many bioactive molecules (Scheme 2.33) [34]. The utility of this new cycloaddition was demonstrated by the excellent regioselectivity obtained using a variety of ynamides and nitriles. For the ynamides bearing various sulfonamide substituents ($R' = n$-butyl, phenyl, and benzyl), the corresponding pyrimidine products were obtained in 75–89% yields. They tested the reactions on the ynamides in which the phenyl group bears different substituents ($X = $ OMe, Cl, and CO_2Me), thus yielding the desired pyrimidines in satisfactory yields. For other heteroaryl-substituted ynamides, the corresponding cycloadducts were produced efficiently. For the 4-substituted benzonitriles bearing electron-donating groups ($X = $ OMe and Me), the gold-catalyzed cycloadditions proceeded smoothly to yield the pyrimidine products in excellent yields. This synthetic method was also feasible for the electron-deficient benzonitriles ($X = CO_2Me$, F, Cl, and Br), thus providing the desired products in 72–84% yields. To their delight, this method was also compatible with the aliphatic nitriles. They postulated that the reaction mechanism involved initial attack of a p-alkyne on two nitriles followed by a ring closure.

Scheme 2.33

A two-neck flask was charged with chloro(triphenylphosphine)AuCl (11.8 mg, 0.0239 mmol) and silver bis(trifluoromethanesulfonyl)imide (9.28 mg, 0.0239 mmol), and to this mixture was added dry DCE (1.0 mL); the resulting mixture was stirred at room temperature for 10 min. To this mixture was added a dry DCE solution (3 mL) of N-methyl-N-(phenylethynyl)methanesulfonamide 1a (100 mg, 0.478 mmol) and benzonitrile (197 mg, 1.913 mmol) dropwise. After stirring at 75°C for 4 h, the reaction mixture was filtered over a short celite bed, concentrated, and eluted through a silica column to give the desired product.

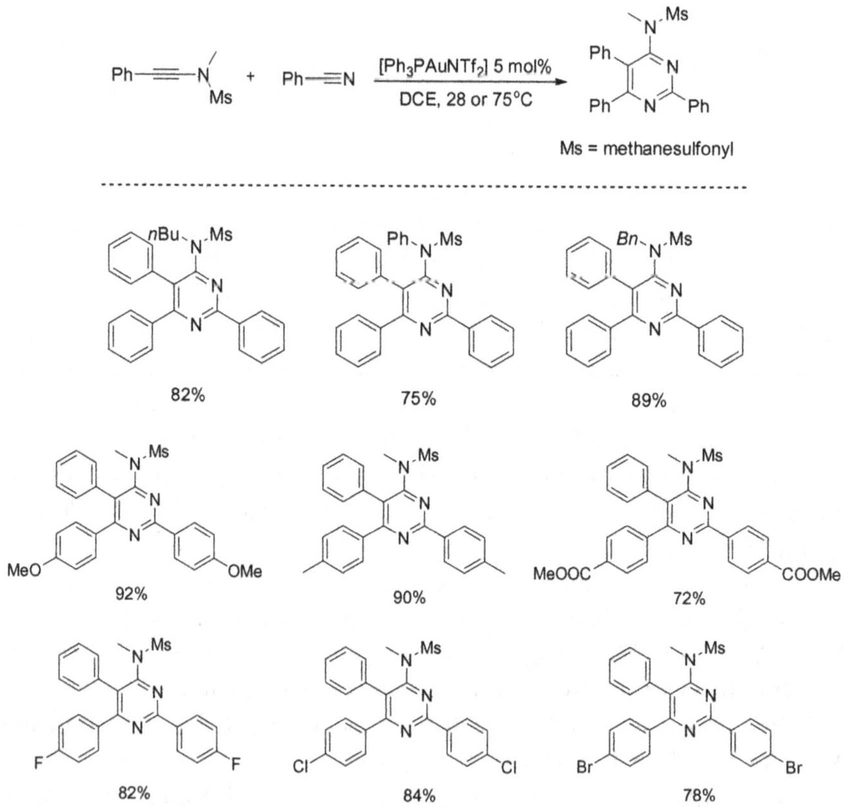

Scheme 2.33 Gold-catalyzed synthesis of polyfunctionalized pyrimidine.

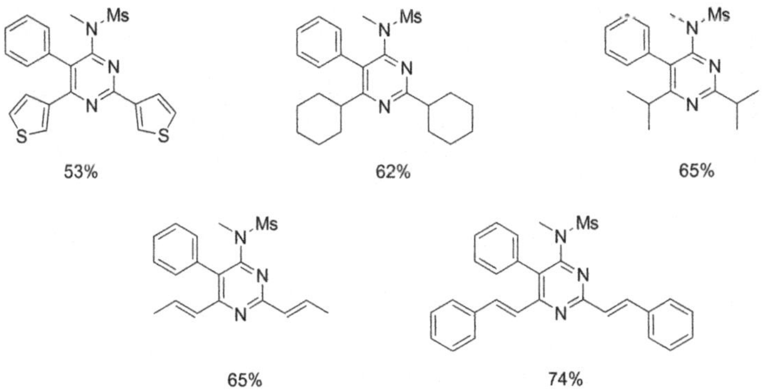

53%   62%   65%

65%   74%

Scheme 2.33 (Continued)

In 2014, Wang and coworkers developed a facile and efficient synthesis of 4-iminopyrimidines via a copper-catalyzed three-component reaction (Scheme 2.34). [35] The reactions were high regioselectivity approached. The reaction tolerated a variety of substituents on imidamides and afforded corresponding 4-iminopyrimidines in yields varying from 62% to 98% without the apparent substitution effect. Benzenesulfonyl azides with groups such as Cl or OMe also gave moderate to good yields. To their surprise, methanesulfonyl azide gave the best yield in comparison with other arenesulfonyl azides. The regioselectivity of this reaction was determined by the NOE technique. The NOE clearly indicated that the reaction was highly regioselective.

Scheme 2.34

To an oven-dried Schlenk tube equipped with a magnetic stirring bar were added sequentially 1-(naphthalene-1-yl)-1,3-butadiyne (0.24 mmol), tosyl azide (0.24 mmol), *N*-phenylbenzimidamide (0.2 mmol) CuCl (0.02 mmol), and dry DCE (1 mL) at room temperature under N_2 atmosphere. Then Et_3N (0.24 mmol) was added slowly via syringe and the mixture was stirred continually for 3 h at room temperature. Upon completion, the solvent was removed in vacuum and the residue was purified by column chromatography on silica gel.

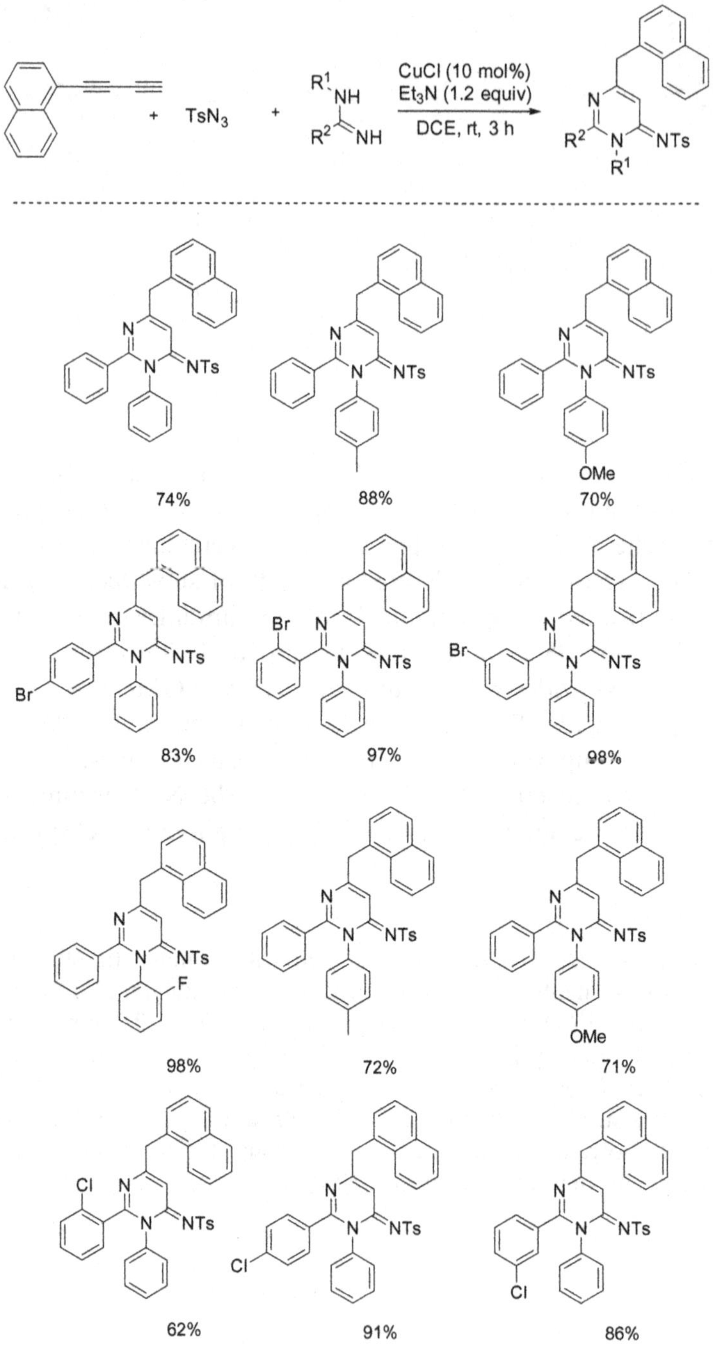

Scheme 2.34 Copper-catalyzed three-component synthesis of pyrimidine.

In 2015, Xiong and coworkers reported a novel tunable regioselective synthesis of pyrazolo[3,4-d]pyrimidine derivatives via aza-Wittig/Ag(I) or base-promoted tandem reaction (Scheme 2.35) [36]. This approach provided a simple and efficient way to construct pyrazolo [3,4-d]-pyrimidine derivatives under mild conditions. And this methodology had been successfully used for the synthesis various different substituted pyrazolo-[3,4-d]pyrimidine derivatives. Various carbodiimides compounds were found to be the suitable reaction partners with methanol or ethanol to provide the corresponding products in moderate to excellent yield. This transformation showed good functional group tolerance. Either electron-withdrawing and electron-donating substituted groups or halogen groups at the aromatic ring could be introduced into the desired product under the standard reaction conditions without any difficulties. It was worth noting that the steric effect had no significant effect to the yields, regardless of the substitution pattern of the aryl ring (*ortho*, *meta*, or *para*) of the aryl isocyanates used in the reaction, the corresponding products were obtained in good yields. The steric and electronic properties of the substituents on the phenyl rings had little effect on the yield. The methyl, methoxyl, ethoxyl, fluorine, and chlorine at the aromatic ring of aryl isocyanates all reacted smoothly affording the desired products in good to excellent yields.

Scheme 2.35

To the solution of iminophos-phorane (0.5 mmol) in dry CH_2Cl_2 (3 mL), ArNCO (0.6 mmol) was added under N_2. The mixture was stirred at 25°C for 12−24 h, then the solvent was removed under vacuum. The residue was dissolved in ROH (4 mL) and NaOH(0.5 mmol, 1.0 equiv) was added and the mixture was stirred at 25°C for 21 h. After the solvent was removed, product was obtained by flash chromatography.

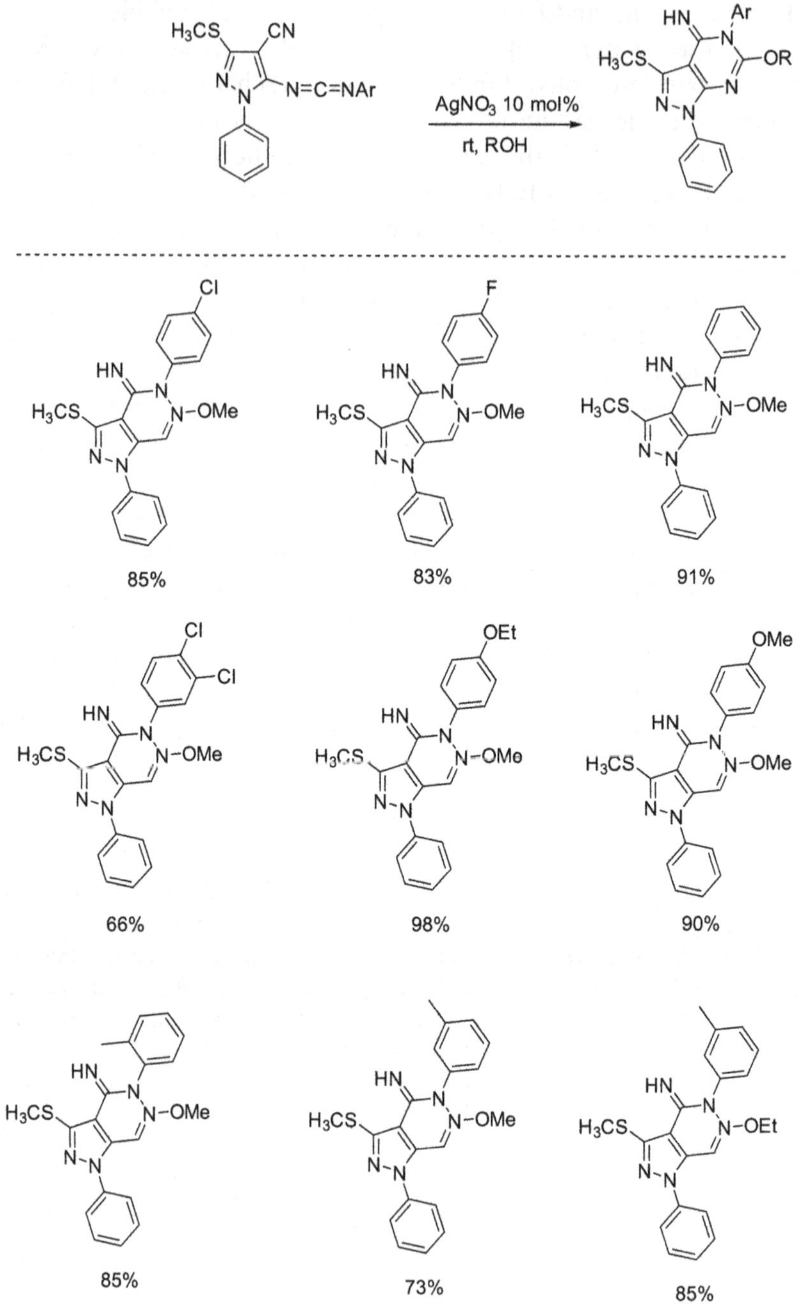

Scheme 2.35 *Ag(I)-catalyzed synthesis of pyrimidine.*

REFERENCES

[1] Ingebrigtsen, T.; Helland, I.; Lejon, T. *Heterocycles* **2005**, *65*, 2593–2603.

[2] Yavari, I.; Nematpour, M.; Ghazanfarpour-Darani, M. *Tetrahedron Lett.* **2012**, *53*, 942–943.

[3] Mahdavi, M.; Kianfard, H.; Saeedi, M.; Ranjbar, P. R.; Shafiee, A. *Synlett.* **2016**, *27*, 1689–1692.

[4] a. Gore, R. P.; Rajput, A. P. *Drug Invent. Today* **2013**, *5*, 148–152.
b. Selvam, T. P.; James, C. R.; Dniandev, P. V.; Valzita, S. K. *Res. Pharm.* **2012**, *2*, 1–13.

[5] Karpov, A. S.; Müller, T. J. J. *Synthesis* **2003**, *18*, 2815–2826.

[6] Karpov, A. S.; Müller, T. J. J. *Org. Lett.* **2003**, *19*, 3451–3454.

[7] Karpov, A. S.; Merkul, E.; Rominger, F.; Muller, T. J. J. *Angew. Chem. Int. Ed.* **2005**, *44*, 6951–6956.

[8] Stonehouse, J. P.; Chekmarev, D. S.; Ivanova, N. V.; Lang, S.; Pairaudeau, G.; Smith, N., et al. *Synlett.* **2008**, *1*, 100–104.

[9] Rostamizadeh, S.; Nojavan, M.; Aryan, R.; Sadeghian, H.; Davoodnejad, M. *Chinese Chem. Lett.* **2013**, *24*, 629–632.

[10] Kumar, A.; Kumar, M.; Maurya, S.; Khanna, R. S. *J. Org. Chem.* **2014**, *79*, 6905–6912.

[11] Deibl, N.; Ament, K.; Kempe, R. *J. Am. Chem. Soc.* **2015**, *137*, 12804–12807.

[12] Müller, T. J. J.; Braun, R.; Ansorge, M. *Org. Lett.* **2000**, *13*, 1967–1970.

[13] Bellura, E.; Langer, P. *Tetrahedron* **2006**, *62*, 5426–5434.

[14] Xie, F.; Li, S.; Bai, D.; Lou, L.; Hu, Y. *J. Comb. Chem.* **2007**, *9*, 12–13.

[15] Bagley, M. C.; Lin, Z.; Pope, S. J. A. *Tetrahedron Lett* **2009**, *50*, 6818–6822.

[16] Lin, M.; Chen, Q.; Zhu, Y.; Chen, X.; Cai, J.; Pan, Y. M., et al. *Synlett* **2011**, *8*, 1179–1183.

[17] Chao, B.; Lin, S.; Ma, Q.; Lu, D.; Hu, Y. *Org. Lett.* **2012**, *9*, 2398–2401.

[18] Nayak, M.; Rastogi, N.; Batra, S. *Eur. J. Org. Chem.* **2012**, 1360–1366.

[19] Zhou, N.; Xie, T.; Li, Z.; Xie, Z. *Chem. Eur. J.* **2014**, *20*, 17311–17314.

[20] Zhan, H.; Chen, L.; Tan, J.; Cao, H. *Catalysis Commun* **2016**, *73*, 109–112.

[21] Zieliński, W.; Kudelko, A. *Monatsh. Chem.* **2000**, *131* 895-599.

[22] Yamamoto, K.; Chen, Y. G.; Buono, F. G. *Org. Lett.* **2005**, *21*, 4673–4676.

[23] Lu, W.; Song, W.; Hong, D.; Lu, P.; Wang, Y. *Adv. Synth. Catal.* **2009**, *351*, 1768–1772.

[24] Sasada, T.; Kobayashi, F.; Sakai, N.; Konakahara, T. *Org. Lett.* **2009**, *10*, 2161–2164.

[25] Li, D.; Duan, S.; Hu, Y. *J. Comb. Chem.* **2010**, *12*, 895–899.

[26] Majumder, S.; Odom, A. L. *Tetrahedron* **2010**, *66*, 3152–3158.

[27] Zhang, Q.; Sun, J.; Zhu, Y.; Zhang, F.; Yu, B. *Angew. Chem. Int. Ed.* **2011**, *50*, 4933–4936.

[28] Han, B.; Han, R.; Ren, Y.-W.; Duan, X.; Xu, Y.; Zhang, W. *Tetrahedron* **2011**, *67*, 5615–5620.

[29] Sasada, T.; Aoki, Y.; Ikeda, R.; Sakai, N.; Konakahara, T. *Chem. Eur. J.* **2011**, *17*, 9385–9394.

[30] Namitharan, K.; Pitchumani, K. *Adv. Synth. Catal.* **2013**, *355*, 93–98.

[31] Satoh, Y.; Yasuda, K.; Obora, Y. *Organometallics* **2012,** *31*, 5235–5238.

[32] Lane, T. K.; Nguyen, M. H.; D'Souza, B. R.; Spahn, N. A.; Louie, J. *Chem. Commun.* **2013,** *49*, 7735–7737.

[33] You, X.; Yu, S.; Liu, Y. *Organometallics* **2013,** *32*, 5273–5276.

[34] Karad, S. N.; Liu, R.-S. *Angew. Chem. Int. Ed.* **2014,** *53*, 9072–9076.

[35] Xing, Y.; Cheng, B.; Wang, J.; Lu, P.; Wang, Y. *Org. Lett.* **2014,** *16*, 4814–4817.

[36] Wang, T.; Xiong, J.; Wang, W.; Li, R.; Tang, X.; Xiong, F. *RSC. Adv.* **2015,** *5*, 19830–19837.

CHAPTER 3

Synthesis of Pyrazine

As early as in 1989, Ohta and coworkers reported a method to synthesize pyrazine derivatives using palladium-catalyzed coupling of 2-chloro-3,6-dialkylpyrazines with protected indoles (Scheme 3.1) [1]. Reactions of 1-tosylindole with chloropyrazine in the presence of catalytic $Pd(PPh_3)_4$ led predominately to 3-heteroaryl indoles in moderate to good yields. Alternatively, under the same conditions, 1-alkyl- and benzylindoles were shown to undergo substitution at C-2, affording the corresponding 2-heteroarylated products in good yields.

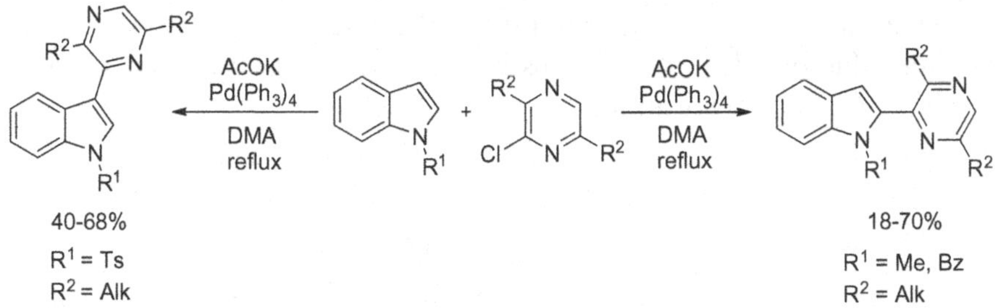
AcOK
Pd(Ph3)4
DMA
reflux
AcOK
Pd(Ph3)4
DMA
reflux
40-68%
R1 = Ts
R2 = Alk
18-70%
R1 = Me, Bz
R2 = Alk

Scheme 3.1 Palladium-catalyzed synthesis of pyrazines using indoles.

Later, Ohta and coworkers demonstrated palladium-catalyzed heteroarylation of other electron-rich heterocycles to synthesize pyrazines (Scheme 3.2) [2]. In this reaction, coupling of chloropyrazines with pyrroles, furans, thiophenes, and a number of azoles in the presence of Pd $(PPh_3)_4$ gave the corresponding pyrazine-substituted products in moderate to very good yields.

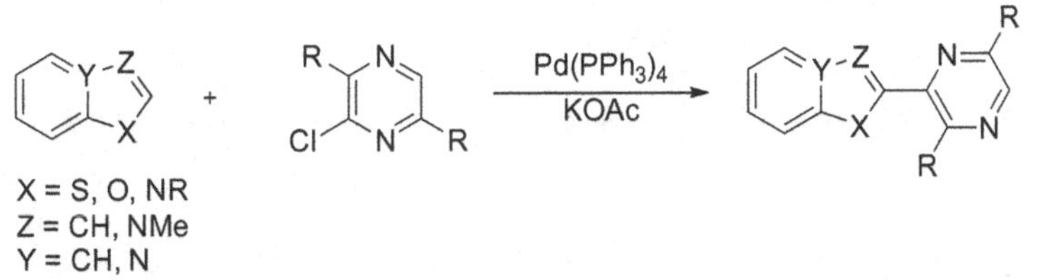
Pd(PPh3)4
KOAc
X = S, O, NR
Z = CH, NMe
Y = CH, N

Scheme 3.2 Palladium-catalyzed synthesis of pyrazines using heterocycles.

Transition Metal Catalyzed Pyrimidine, Pyrazine, Pyridazine and Triazine Synthesis.
DOI: http://dx.doi.org/10.1016/B978-0-12-809378-8.00003-1

In 1996, Sato and coworkers developed syntheses of trisubstituted and tetrasubstituted pyrazines, which gave moderate to good yields. First, they described the synthesis of trialkylpyrazines having methyl groups on C-2 and C-5, using the cross-coupling reaction of 2-chloro-3,6-dimethylpyrazine with dialkylzinc in the presence of [1,3-bis(diphenylphosphino)propane] nickel(II) chloride (Scheme 3.3A) [3]. Generally, the reaction could work at room temperature. However, when chloropyrazines with more bulky alkyl groups were employed, the coupling reaction did not ensue at room temperature. The temperature should be increased to 40−50°C to provide the corresponding trialkylpyrazines in excellent yields. The dimethyl products were acylated with an α-keto acid under the Minisci radical conditions providing 2-acyl-5-alkyl-3,6-dimethylpyrazines (Scheme 3.3B). Higher yields were obtained of isovalerylpyrazines than of the propionyl homologues and the yields probably reflect the stability of acyl radicals. Alkylation competing with acylation was observed in the radical reaction of various heteroaromatics with isobutyraldehyde. The acylation is influenced by steric factors due to the substituent adjacent to the ring carbon available for the substitution.

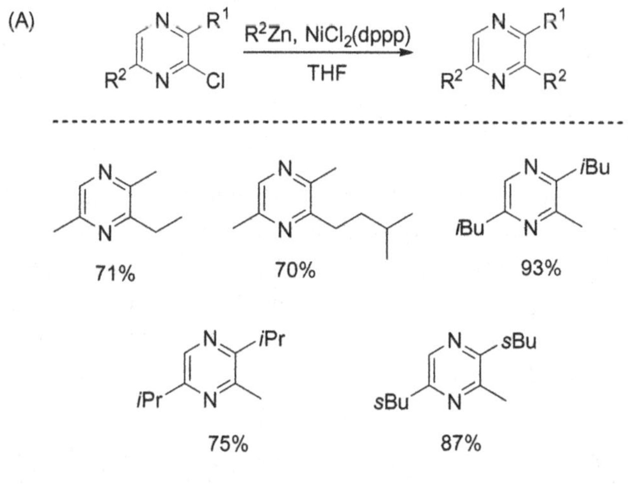

Scheme 3.3 (A) Nickel-catalyzed synthesis of trisubstituted pyrazines. (B) Synthesis of tetrasubstituted pyrazines.

(B)

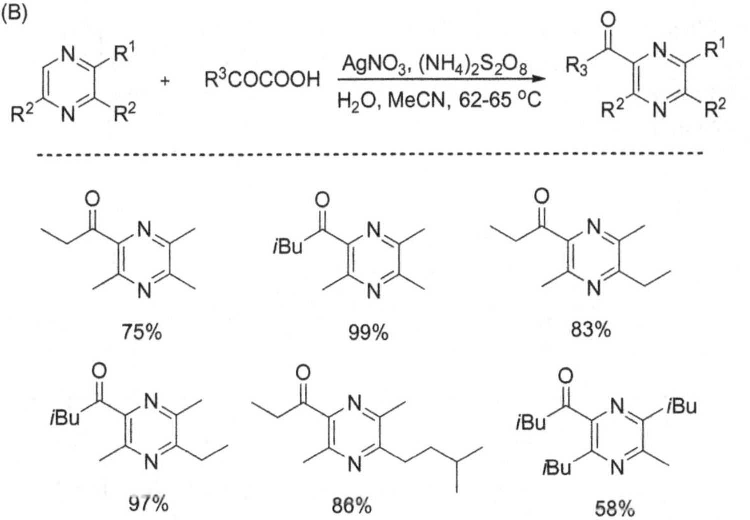

Scheme 3.3 (Continued)

In 2002, Albaneze-Walker and coworkers presented that pyrazines could be decyanated by hydrogenation with platinum on carbon in the presence of activated carbon under acidic conditions (Scheme 3.4) [4]. Pyrazine carbonitrile-N-oxides undergo a stepwise reduction to the deoxy-pyrazinecarbonitriles followed by decyanation to give pyrazines in good yields. Several pyrazine substrates were synthesized and subjected to the optimized conditions. N-Acylated and subjected to the decyanation conditions: acylation of the amine did not hinder the reaction. Only the deoxygenated product was isolated. It may be that this substrate was unable to adhere to the catalyst due to the steric hindrance of the camphor methyl groups.

Scheme 3.4

A slurry of pyrazine N-oxide (2.00 g, 9.4 mmol), 5% Pt/C (0.30 g, 15 mol%), Darco KB-B (1.50 g, 75 mol%), and glacial acetic acid (0.650 mL, 11.3 mmol) in methanol (40 mL) was submitted for hydrogenation at 40 psi and 40°C for 18 h in pressure apparatus. After the hydrogen atmosphere was flushed with nitrogen, a solution of sodium hydroxide (0.450 g, 11.3 mmol) in 2 mL water was added. The reaction mixture was filtered through Solka-Floc to remove the catalyst. Solka Floc/catalyst was rinsed with 1 L methanol. The filtrate contains the majority of cyanide as NaCN.

The filtrate was concentrated to 5–6 mL by a rotary evaporator in a fume hood. Water (10 mL) was added and it was aged for 1 h. The precipitate was collected by filtration and air-dried overnight in a hood to give the product.

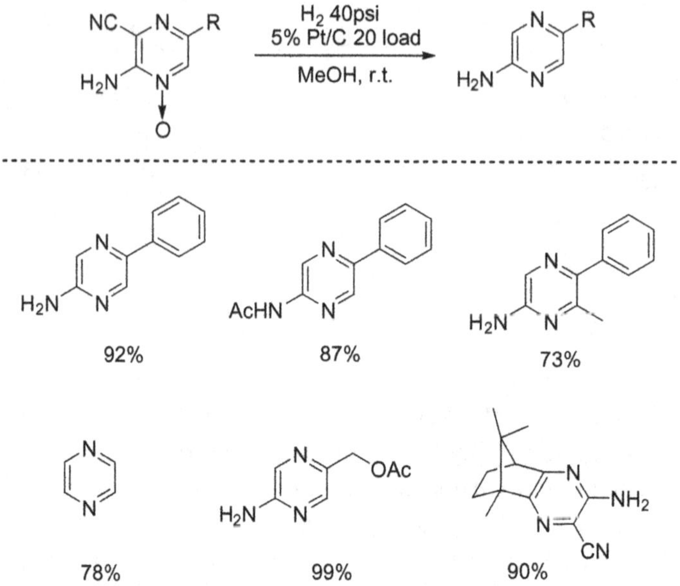

Scheme 3.4 *Pt/C-catalyzed synthesis of pyrazines.*

In 2004, Janda and coworkers developed a new methodology for the efficient synthesis of pyrazin-6-ones and pyrazines utilizing a rhodium-catalyzed N—H insertion reaction between Boc amino acid amides and R-diazo-α-ketoesters as the key step (Scheme 3.5) [5]. These N—H insertion products are easily converted into the corresponding pyrazin-6-ones by acid-promoted cyclodehydration and can be further decorated by N-alkylation with alkyl halides. Alternatively, the pyrazin-6-ones are converted readily into the corresponding 6-bromopyrazines that can be further elaborated using Suzuki cross-coupling reaction with aryl boronic acids.

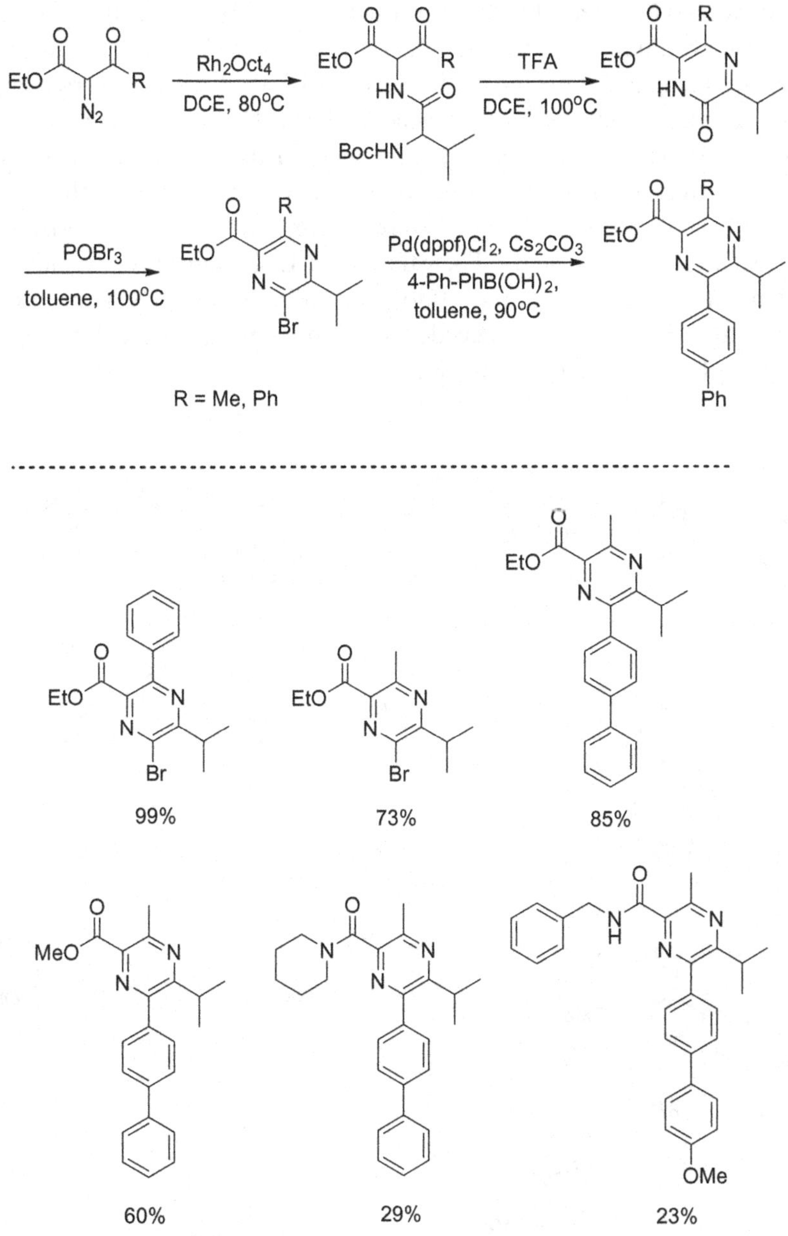

Scheme 3.5 Synthesis of fully aromatized pyrazines.

In 2008, a novel and convenient procedure for the synthesis of asymmetrically tri- and tetrasubstituted pyrazines starting from para-methoxybenzyl-protected 3,5-dichloro-2(1H)-pyrazinones was elaborated by Eycken and coworkers [6]. They used 5-halo-2-(methylthio)-pyrazines which were synthesized by their group as the substrates in the reaction. First, they evaluated the Suzuki-Miyaura cross-coupling and synthesized pyrazines using aryl boronic acids (Scheme 3.6A). A variety of pyrazines were obtained in moderate to excellent yields. However, when the aryl boronic acid was reacted with CF_3, the yield of the product pyrazine was low. Next they investigated the Sonogoshira cross-coupling for these substrates (Scheme 3.6B). Reactions were performed well and the products were isolated in good yields. Then they used Liebeskind-Srogl conditions to generate pyrazines without thiomethylether groups (Scheme 3.6C). Suprisingly, the reactions proceeded smoothly delivering the compounds in excellent yields ranging from 78 to 94%.

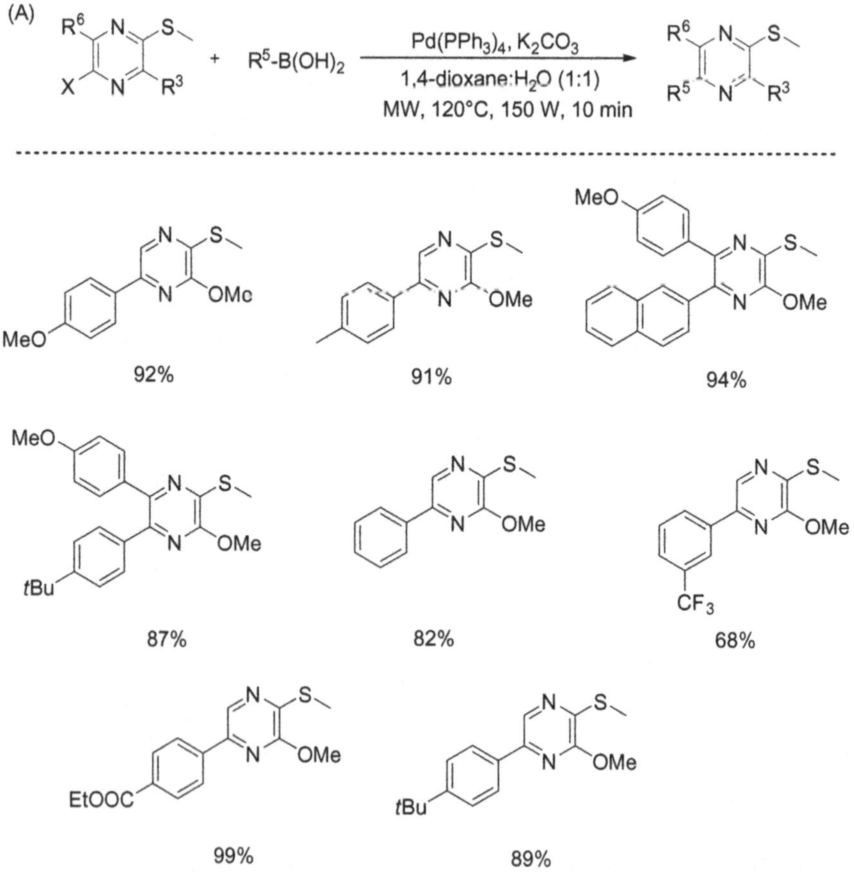

Scheme 3.6 *(A) Microwave-assisted Suzuki-Miyaura coupling. (B) Microwave-assisted Sonogashira reaction. (C) Microwave-assisted Liebeskind-Srogl coupling.*

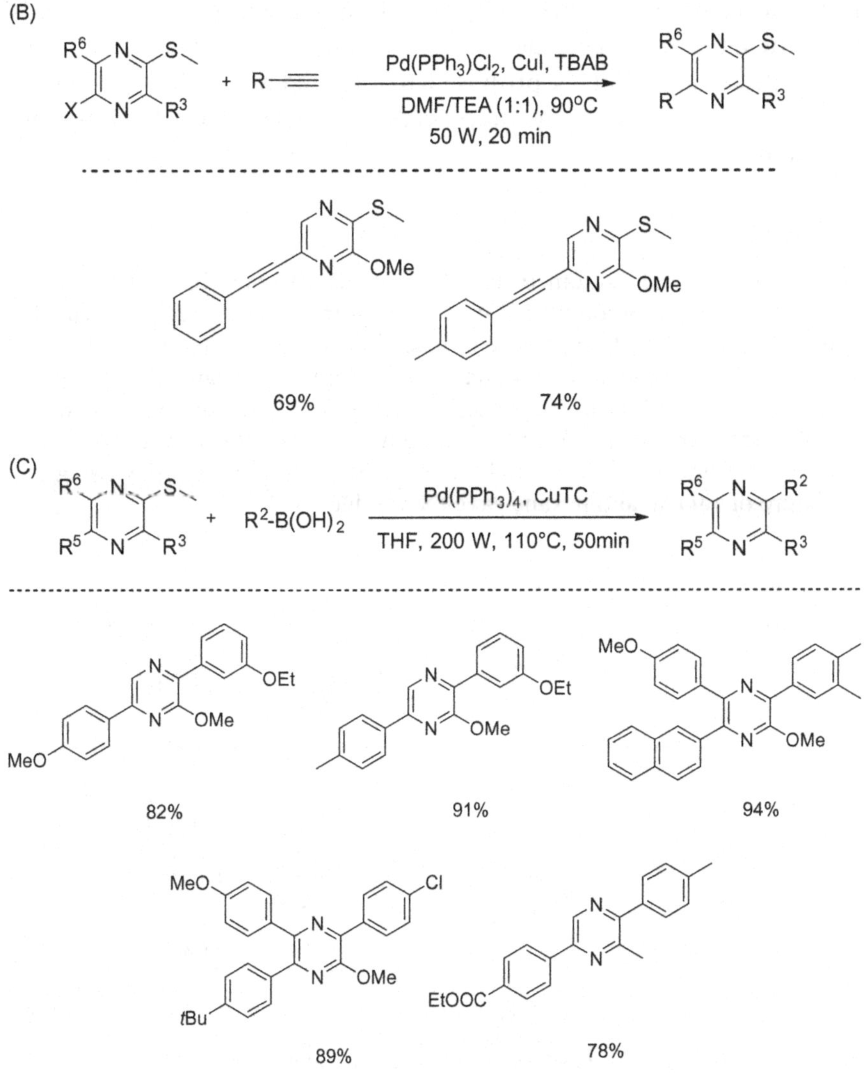

(B)

69%   74%

(C)

82%   91%   94%

89%   78%

Scheme 3.6 Continued.

In 2008, Kang's group developed palladium-catalyzed direct arylation of tautomerizable heterocycles with aryl boronic acids using phosphonium salts to synthesize pyrazines (Scheme 3.7) [7]. Complete conversion and good to excellent isolated yields were observed for all the heterocycles employed. Both electron-rich and electron-poor aryl boronic acids coupled well with the heterocycle-phosphonium salt to afford the pyrazines in excellent yields. Direct arylation using heteroaryl and sterically hindered aryl boronic acids also efficiently furnished the products in high yields. The mechanism of the direct arylation is proposed to proceed via a domino seven-step process including the unprecedented heterocycle-Pd(II)-phosphonium species.

Scheme 3.7

A mixture of the tautomerizable heterocycle (0.5 mmol), PyBroP (1.2 equiv) and Et$_3$N (3 equiv) in 1,4-dioxane (4 mL) was stirred in a sealed tube at rt for 2 h. Then, the aryl boronic acid (2 eq), Na$_2$CO$_3$ (5 eq), PdCl$_2$(PPh$_3$)$_2$ (5 mol%) and water (1 mL) were added, and the mixture was stirred at 100°C in the sealed tube for 4 h. After the mixture was cooled to romm temperature, it was diluted with EtOAc, washed with water, and dried with brine and Na$_2$SO$_4$. Flash chromatography using a mixture of EtOAc and hexane gave the product.

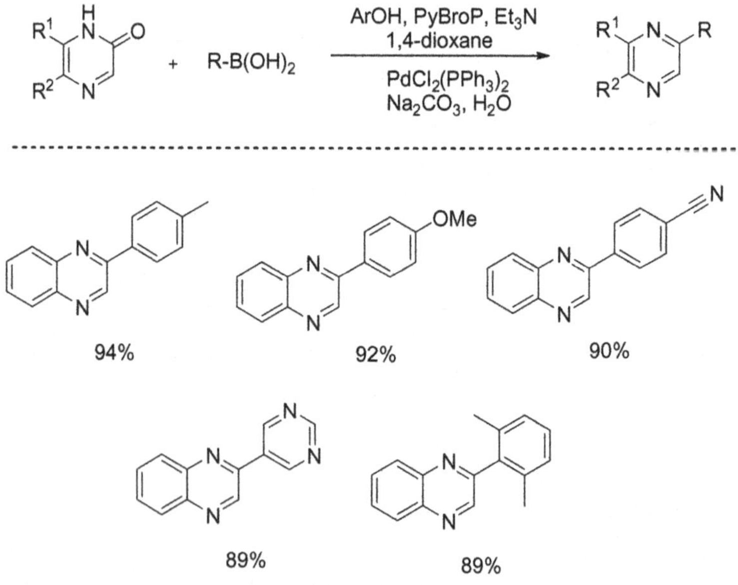

Scheme 3.7 Pd-catalyzed direct arylation of heterocycles to synthesize pyrazines.

In 2010, Eycken's group reported an expedient route for the synthesis of differently substituted pyrazines applying a microwave-assisted palladium-catalyzed phosphonium coupling procedure (Scheme 3.8A) [8]. In this reaction, an array of substituted 2(^{1}H)-pyrazinones were reacted with various aryl or (hetero)aryl boronic acids. For the substrates bearing a methoxy or amino butyl group at the C3-position, the corresponding C2-monosubstituted product was predominantly formed. However, substrates bearing an -SPh or -NH$_2$ at the C3-position were less successful by applying these conditions. Then, they evaluated the phosphonium-mediated Sonogashira-Hagihara-type cross-coupling of 5-chloro-3-methoxy-6-methylpyrazin-2(1H)-one (Scheme 3.8B). A mixture of pyrazinone, PyBroP and Et$_3$N in 1,4-dioxane was irradiated at a ceiling temperature of 65°C and a maximum power of 20 W for 25 min. Next, Pd(PPh$_3$)$_2$Cl$_2$, CuI, and p-tolyl acetylene in DMF (1 mL) were added, and the mixture was irradiated at a ceiling temperature of 85°C and a maximum power of 15W for 25 min. The corresponding mono-alkynylated product was obtained in 69% yield.

Scheme 3.8

To a 10 mL oven-dried microwave vial were added pyrazinone (44 mg, 0.25 mmol), PyBroP (128 mg, 0.275 mmol), Et$_3$N (67 μL, 0.5 mmol), and 1,4-dioxane (2 mL). The reaction tube was sealed and irradiated at a ceiling temperature of 65°C using 10 W maximum power for 25 min. The reaction mixture was cooled with an air flow for 3 min, and Pd (PPh$_3$)$_2$Cl$_2$ (8.8 mg, 5 mol%), boronic acid (36 mg, 0.262 mmol), Na$_2$CO$_3$ (53 mg, 0.5 mmol), and H$_2$O (1 mL) were successively added. The reaction tube was sealed again and irradiated at a ceiling temperature of 85°C using 15 W maximum power for 30 min. The reaction mixture as cooled with an air flow. The mixture was washed with brine (50 mL) and extracted with dichloromethane (2 × 50 mL). The organic layer was dried over MgSO$_4$, and the solvent was removed under reduced pressure. Flash chromatography using a mixture of EtOAc and hexane gave the product.

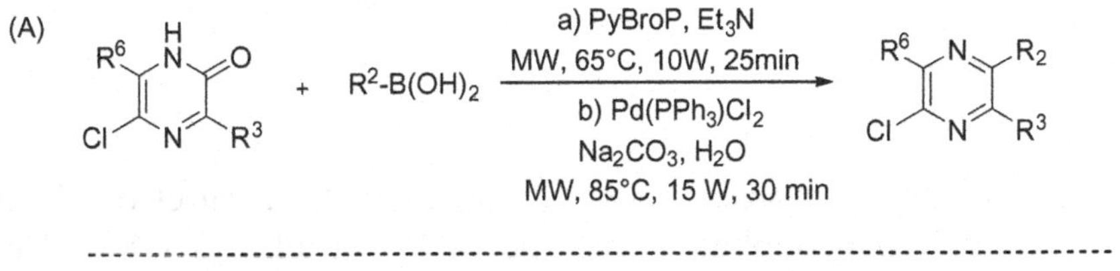

Scheme 3.8 *(A) Microwave-assisted palladium-catalyzed synthesis of pyrazines using (hetero)aryl boronic acids. (B) Microwave-assisted palladium-catalyzed synthesis of pyrazines using alkynes.*

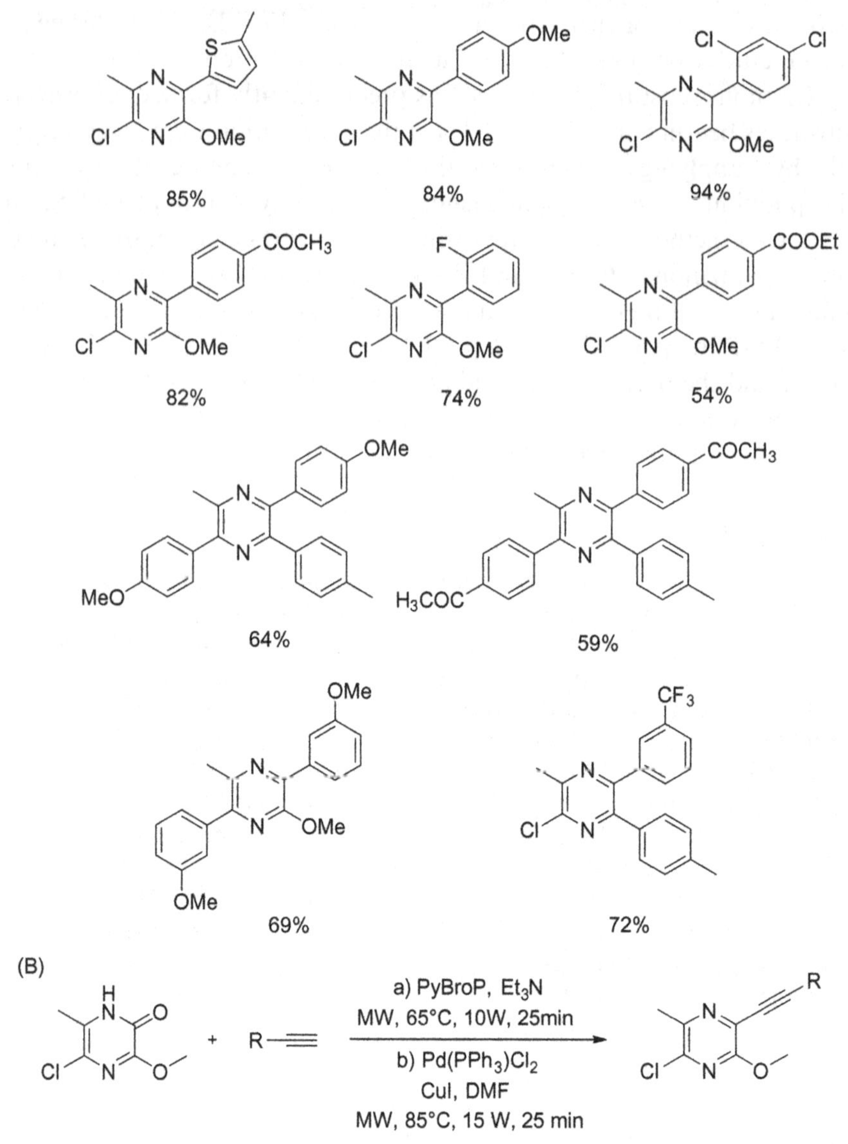

85%

84%

94%

82%

74%

54%

64%

59%

69%

72%

(B)

a) PyBroP, Et₃N
MW, 65°C, 10W, 25min

b) Pd(PPh₃)Cl₂
CuI, DMF
MW, 85°C, 15 W, 25 min

Scheme 3.8 (Continued)

In 2010, Kang's group reported the first chemoselective direct dehydrative cross-coupling of heterocycles with alkynes via Pd/Cu-catalyzed phosphonium coupling to synthesize pyrazines (Scheme 3.9) [9]. They found that aryl alkynes, alkenyl alkynes, and alkyl alkynes all worked well in moderate to high yields at room temperature under ambient atmosphere using the Pd/Cu catalyzed condition. However, in the case of an electron-deficient alkyne, the Cu-free condition was found to give a higher yield. this new Pd/Cu-catalyzed phosphonium coupling was applied to other tautomeriz-able heterocycles with p-tolylacetylene. We found that the coupling reactions of these tautomerizable heterocycles appeared to be some-what slower under the same conditions, probably due to their slightly different reactivity. We found that these reactions can be accelerated under the modified conditions using a slightly stronger base (iPr$_2$NEt) at higher temperatures (50–80°C).

Scheme 3.9

After the reaction mixture of the tautomerizable heterocycle (0.5 mmol), PyBroP (1.2 equiv) and ^{i}Pr$_2$NEt (6 equiv) in 1,4-dioxane (5 mL) was stirred at 50°C under ambient atmosphere in a capped vial for 2 h, p-tolylacetylene (2 equiv), PdCl$_2$(PPh$_3$)$_2$ (5 mol%), and CuI (10 mol%) were added, and it was stirred at 80°C for 18 h. Then, it was diluted with EtOAc, washed with water and brine, and dried over Na$_2$SO$_4$. Flash chromatography using a mixture of EtOAc and hexane gave the coupling product.

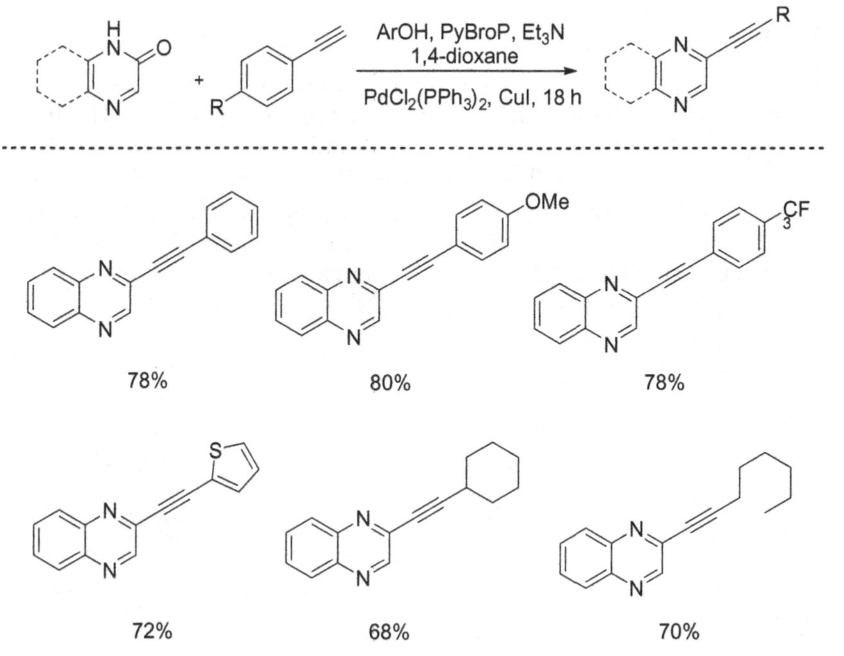

Scheme 3.9 Pd/Cu-catalyzed synthesis of pyrazines.

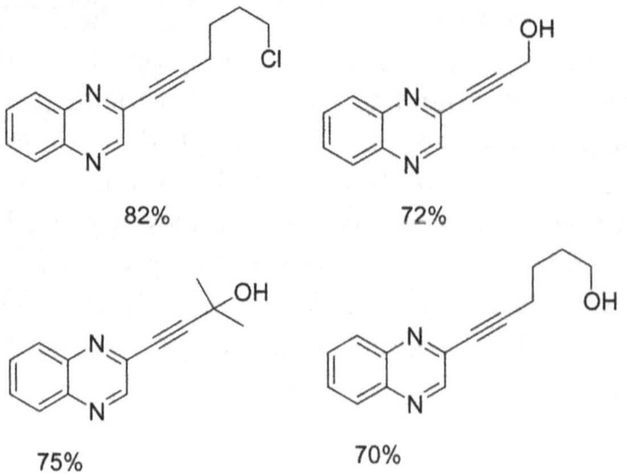

82%   72%

75%   70%

Scheme 3.9 (Continued)

In 2011, Milstein developed a new method to synthesize pyrazines using Ru complex as the catalyst (Scheme 3.10) [10]. These unprecedented reactions proceed under neutral reaction conditions and generate no waste, thereby representing a clear departure from traditional synthetic methodology.

Scheme 3.10

Ru complex (0.02 mmol), isoleucinol (2 mmol) and toluene (2 mL) were added to flask under an atmosphere of nitrogen in a glovebox. The flask was equipped with a condenser and the solution was vigorously refluxed at 165°C with stirring without protection from air for 24 h. The reaction products were analyzed by GC-MS. The reaction mixture was evaporated and the crude product was subjected to silica-gel column chromatography using EtOAc:hexane to afford the corespongding product.

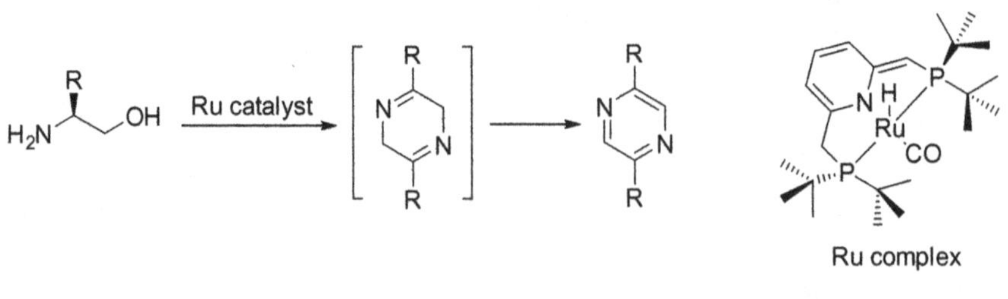

Scheme 3.10 Synthesis of pyrazines using Ru complex catalyst.

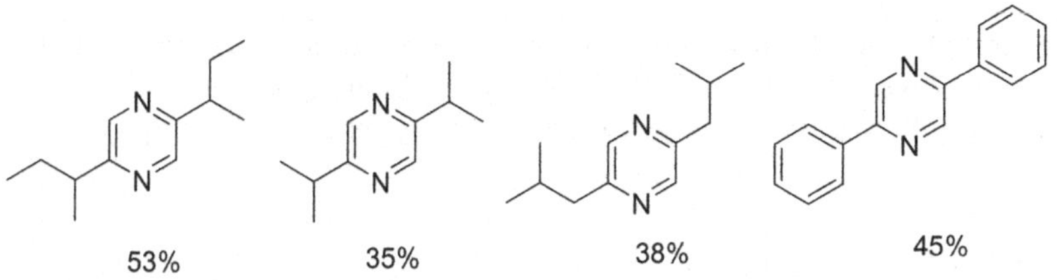

53%   35%   38%   45%

Scheme 3.10 (Continued)

In 2012, Clark and coworkers described a regioselective synthesis of pyrazine derivatives using a palladium-catalyzed amide coupling reaction, which provided quick access to products with substitution at N1 and C2. (Scheme 3.11) [11]. Pyrazine derivatives could be obtained in moderate to good yields. They ascribed this rate enhancement to faster oxidative addition with the electron-deficient pyrazine moiety. In addition, they have shown the potential for further functionalization by installation of primary, secondary, and tertiary amines at the 2-position.

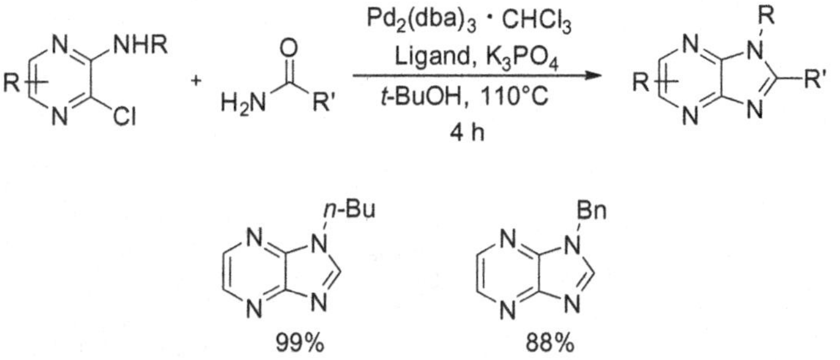

99%   88%

Scheme 3.11 Palladium-catalyzed synthesis of pyrazine derivatives.

Scheme 3.12

To a solution of (hetero)arene (1 mmol) in dichloromethane (8 mL) was added trifluoroacetic acid (80 μL, 1 mmol) followed by arylboronic acid (1.1 mmol). Water (8 mL) was then added, followed by iron(II) acetylacetonate (0.2 mmol) and potassium persulfate (3 mmol). TBAB was then added [5 mol% with respect to (hetero)arenes], and the solution was stirred vigorously at room temperature until completion as monitored by TLC. Then reaction mixture was diluted with dichloromethane and washed with saturated sodium bicarbonate solution. The layers were separated, and the aqueous layer was extracted with dichloromethane. Organic layers were compiled, dried over sodium sulfate, and evaporated in vacuo. Purification was performed by silica gel chromatography.

In 2013, Singh and coworkers reported an iron-catalyzed cross-coupling reaction to synthesize pyrazine derivatives. In this article, iron(II) acetylacetonate along with oxidant ($K_2S_2O_8$) and phase-transfer catalyst (TBAB) under open flask conditions efficiently catalyzed the cross-coupling of pyrazine with arylboronic acids and gave monoarylated products in good to excellent yields (Scheme 3.12) [12]. Various ortho-, meta-, and parasubstituted organoboronic acids were used and provided varying yields. Arylboronic acids possessing electron-donating groups at the para, meta, and ortho position smoothly underwent cross-coupling reaction and gave good to excellent yields. Similarly, arylboronic acids with electron-withdrawing groups gave comparatively lower yields.

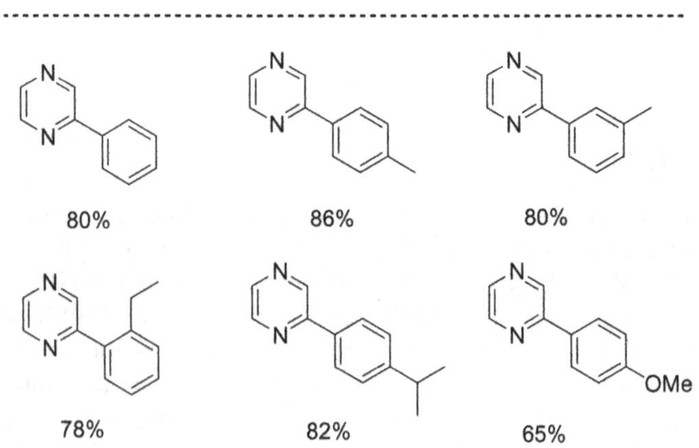

Scheme 3.12 Iron-catalyzed synthesis of pyrazines.

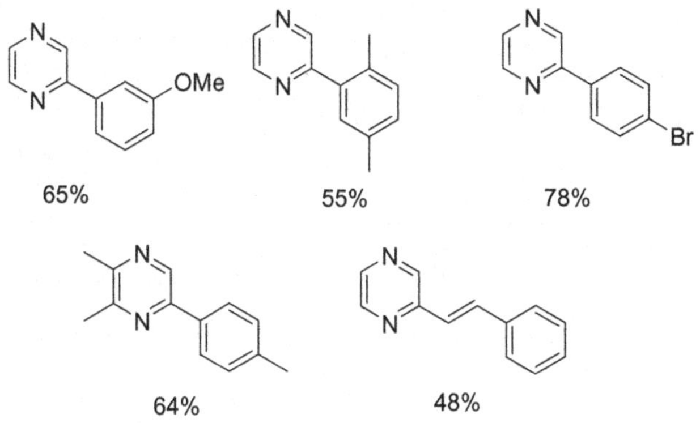

65%   55%   78%

64%   48%

Scheme 3.12 (Continued)

In 2015, an efficient synthetic route to a wide range of trisubstituted pyrazines was presented by Lee and coworkers, which was developed from Rh-catalyzed reaction of 2*H*-azirines with *N*-sulfonyl-1,2,3-triazoles through the elimination of nitrogen molecule and arylsulfinic acid (Scheme 3.13) [13]. In this reaction, electronic modification of substituents did not largely influence efficiency of the reaction. The substrates of 1,2,3-triazoles with electron-donating 3-methyl, 4-methyl, 2-methoxy, and 3-methoxy groups on the phenyl ring underwent the Rh-catalyzed reactions, affording the desired pyrazines in good yields ranging from 70 to 74%. And the reactions of substrates with electron-withdrawing 3-chloro, 4-chloro, 3-bromo, and 4-bromo groups on the phenyl ring provided the cyclization products in moderate to good yields ranging from 51 to 81%.

Scheme 3.13

To a screw-top V-vial were added 2H-azirine derivatives (0.2 mmol), triazole derivatives (0.3 mmol), and $Rh_2(Oct)_4$ (3.1 mg, 0.004 mmol) in EtOAc (1.0 mL). The resulting mixture was stirred at 120°C for 16 h. After Celite filtration and evaporation of the solvents in vacuo, the crude product was purified by column chromatography on silica gel (EtOAc: Hx = 1:5).

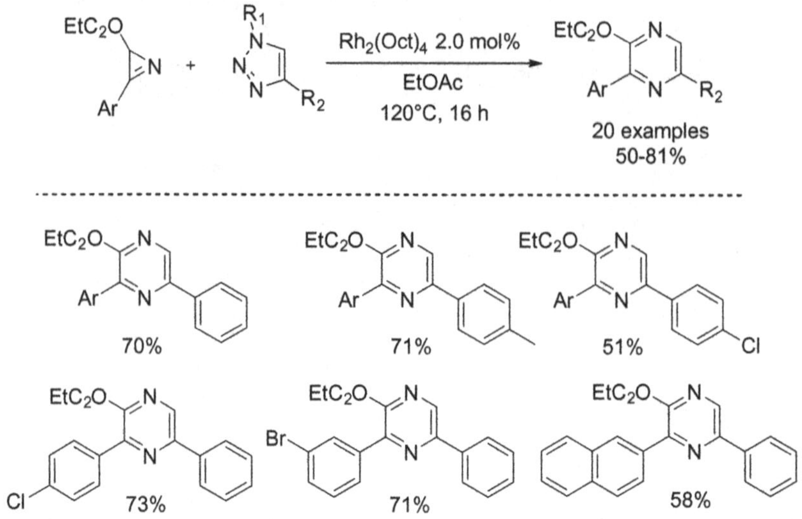

Scheme 3.13 Synthesis of trisubstituted pyrazines.

In 2015, Park and coworkers reported the synthesis of unsymmetrical pyrazines (Scheme 3.14) [14]. α-Diazo oxime ethers have been demonstrated to serve as an excellent precursor for α-imino carbenoids, of which their synthetic utility has been explored in the synthesis of various N-heterocycles. The transformation is generally well tolerated and provides the corresponding pyrazines in good yields. Electronic effect of 2H-azirines on the reaction was examined by substitution of the phenyl group of 2H-azirine, revealing broad tolerance for both electron-donating and -withdrawing groups. The reaction with 2H-azirine bearing ortho-substitution also smoothly proceeded to give the desired product indicating marginal steric influence. Additionally, when the reaction was performed in a gram scale, it also proceeded smoothly to give the corresponding pyrazines in high yields.

Scheme 3.14

Into an oven-dried sealed tube was added 2 mol % of Cu(hfacac)$_2$ in DCE (0.5 mL), followed by the addition of azirine (0.3 mmol, 1 equiv) in DCE (0.5 mL), the reaction was allowed to stir at room temperature with the dropwise addition of α-diazo oxime ether (1.2 equiv) in DCE (1 mL). The reaction was heated at 90°C until the consumption of the azirine, as observed by TLC. The reaction then was further heated at 150°C approximately 10 h. Upon completion of the reaction, the crude was concentrated and purified by flash column chromatography (2:1 Hexane: Ethyl acetate) to give pyrazine.

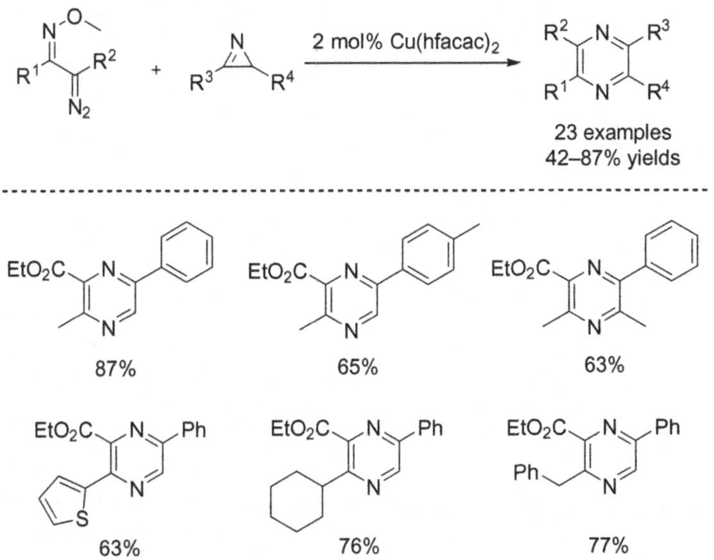

Scheme 3.14 Cu-catalyzed synthesis of pyrazines.

REFERENCES

[1] Akita, Y.; Itagaki, Y.; Takizawa, S.; Ohta, A. *Chem. Pharm. Bull.* **1989,** *37,* 1477–1480.

[2] Aoyagi, Y.; Inoue, A.; Koizumi, I.; Hashimoto, R.; Tokunaga, K.; Gohma, K., et al. *Heterocycles* **1992,** *33,* 257–272.

[3] Sato, N.; Matsuura, T. *J. Chem. Soc. Perkin Trans.* **1996,** *1,* 2345–2350.

[4] Albaneze-Walker, J.; Zhao, M.; Baker, M. D.; Dormer, P. G.; McNamara, J. *Tetrahedron Lett* **2002,** *43,* 6747–6750.

[5] Matsushita, H.; Lee, S.; Yoshida, K.; Clapham, B.; Koch, G.; Zimmermann, J., et al. *Org. Lett.* **2004,** *6,* 4627–4629.

[6] Mehta, V. P.; Sharma, A.; Hecke, K. V.; Meervelt, L. V.; Eycken, E. V. *J. Org. Chem.* **2008,** *73,* 2382–2388.

[7] Kang, F.; Sui, Z.; Murray, W. V. *J. Am. Chem. Soc.* **2008,** *130,* 11300–11302.

[8] Mehta, V. P.; Modha, S. G.; Eycken, E. V. V. *J. Org. Chem.* **2010,** *75,* 976–979.

[9] Kang, F.; Lanter, J. C.; Cai, C.; Sui, Z.; Murray, W. V. *Chem. Commun.* **2010,** *46,* 1347–1349.

[10] Gnanaprakasam, B.; Balaraman, E.; Yehoshoa, B. D.; Milstein, D. *Angew. Chem. Int. Ed.* **2011,** *50,* 12240–12244.

[11] Rosenberg, A. J.; Zhao, J.; Clark, D. A. *Org. Lett.* **2012,** *14,* 1764–1767.

[12] Singh, P. P.; Aithagani, S. K.; Yadav, M.; Singh, V. P.; Vishwakarma, R. A. *J. Org. Chem.* **2013,** *78,* 2639–2648.

[13] Ryu, T.; Baek, Y.; Lee, P. H. *J. Org. Chem.* **2015,** *80,* 2376–2383.

[14] Loy, N. S. Y.; Kim, S.; Park, C. *Org. Lett.* **2015,** *17,* 395–397.

CHAPTER 4

Synthesis of Pyridazine

In 2009, Williams reported a method to synthesize pyridazines by a ruthenium-catalyzed isomerization of alkynediols and in situ cyclization (Scheme 4.1) [1]. In this reaction, alkyne-1,4-diols are readily available substrates which are isomerized to 1,4-diketones using $Ru(PPh_3)_3(CO)H_2$/xantphos as a catalyst. They proved that the reaction was unsuccessful if the hydrazine was added at the start of the reaction, as the isomerization step was inhibited, and hence hydrazine was added three hours after the start of the reaction. And they found that the most successful reaction conditions involved either the addition of styrene as a hydrogen acceptor or activation of the catalyst using KOt-Bu. Besides, crotononitrile inhibited the isomerization step. Cyclohexene and hexene provided a mixture of products with minimal pyridazine formation. When they increased the amount of base, the isomerization was incomplete. Surprisingly, the combined use of styrene and KOt-Bu led to a complex mixture of products. Whilst there was complete consumption of starting material and no remaining diketone, several minor impurities were observed.

Scheme 4.1

To an oven-dried, argon-purged carousel tube was added $Ru(PPh_3)_3(CO)H_2$ (23 mg, 0.025 mmol) and xantphos (14 mg, 0.025 mmol). Degassed anhydrous toluene (1 mL) was added and the reaction heated to reflux for 1 h. The reaction was cooled to room temperature before the substrate was added and the reaction returned to reflux for 3 h. The reaction was cooled to room temperature and hydrazine hydrate (0.049 mL, 1 mmol) was added. The reaction was heated at reflux for 17 h before cooling and addition of styrene (0.12 mL, 1 mmol). The reaction was then heated at reflux for a further 4 h. The solvent was removed in vacuo and the pyridazine removed from the ruthenium residue by extracting with hexane.

Transition Metal Catalyzed Pyrimidine, Pyrazine, Pyridazine and Triazine Synthesis.
DOI: http://dx.doi.org/10.1016/B978-0-12-809378-8.00004-3

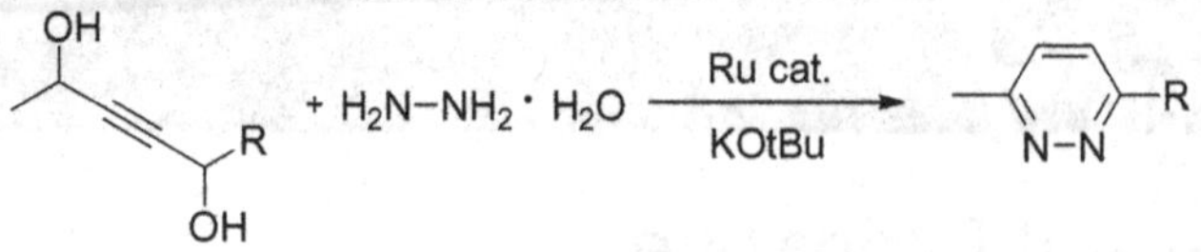

Scheme 4.1 Ruthenium-catalyzed synthesis of pyridazines.

In 2011, Yamane developed a simple and efficient synthesis of pyridazines that was achieved by a palladium-catalyzed annulation methodology, which gave products in moderate to good yield (Scheme 4.2) [2]. Several internal alkynes were applicable to this reaction and it was compatible with a number of functional groups such as such as alkyl, methoxy, cyano, nitro, trifluoromethyl, acetyl, and methoxycarbonyl groups. Symmetrical alkynes were used to check the functional group tolerance and the products were obtained in good yields. Besides, carbonyl and chloro groups on the phenyl ring of the internal alkyne were found to be tolerated. The consumption of the iodophenyltriazene was faster in the reaction with an electron-deficient alkyne, such as ethyl phenylpropiolate. They anticipated coordination of the ethoxy group of the alkyne may facilitate the regioselective addition of organopalladium intermediate.

Scheme 4.2

A solution of compound 1 (0.25 mmol), alkyne (0.75 mmol), $PdCl_2$ (3.3 mg, 0.02 mmol), $P(o\text{-}Tolyl)_3$ (11.4 mg, 0.04 mmol), and nBu_3N (119 mL, 0.50 mmol) in DMF (5 mL) was stirred at 90°C. The reaction mixture was allowed to cool to room temperature. The solvent was evaporated and the residue was purified by column chromatography on silica gel (hexane/ethyl acetate/dichloromethane).

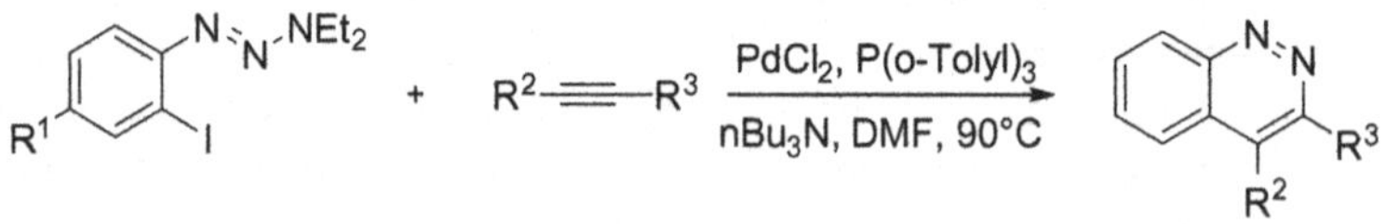

Scheme 4.2 Palladium-catalyzed synthesis of pyridazines.

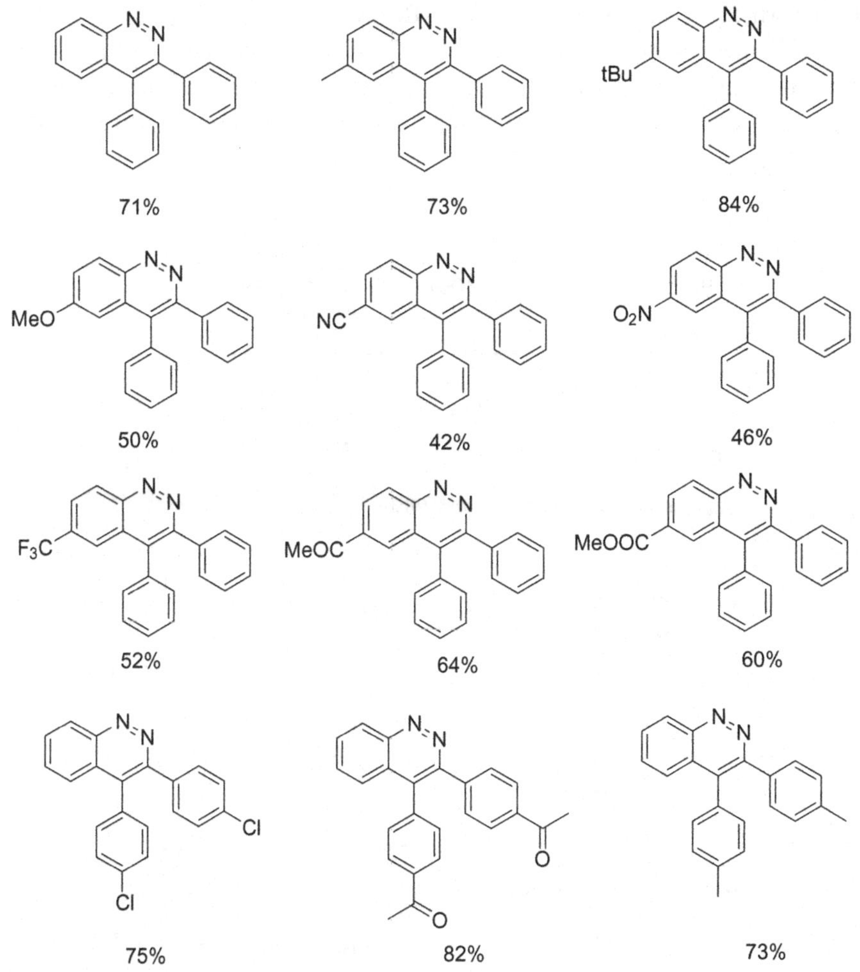

71%

73%

84%

50%

42%

46%

52%

64%

60%

75%

82%

73%

Scheme 4.2 (Continued)

In 2012, Willis and coworkers demonstrated a method to perform a two-step route to an underused class of heterocyclic compounds, the pyridazines (Scheme 4.3) [3]. In this reaction, the use of readily available starting materials and catalysts, and the ability to introduce functionality at each position of the ring were the superiorities. A number of heterocycle-based substrates could be used to tolerate the reaction conditions. Presumably because of their increased propensity for oxidation, they could also synthesize pyridazines with strongly electron-donating substituents. However, less electron-rich substrates were transformed with lower efficiency. Besides, the two-step route could be performed using a one-pot synthesis without isolation of any intermediate. Instead, the crude reaction mixture from the Cu-catalyzed transformation was simply filtered through celite, concentrated under vacuum.

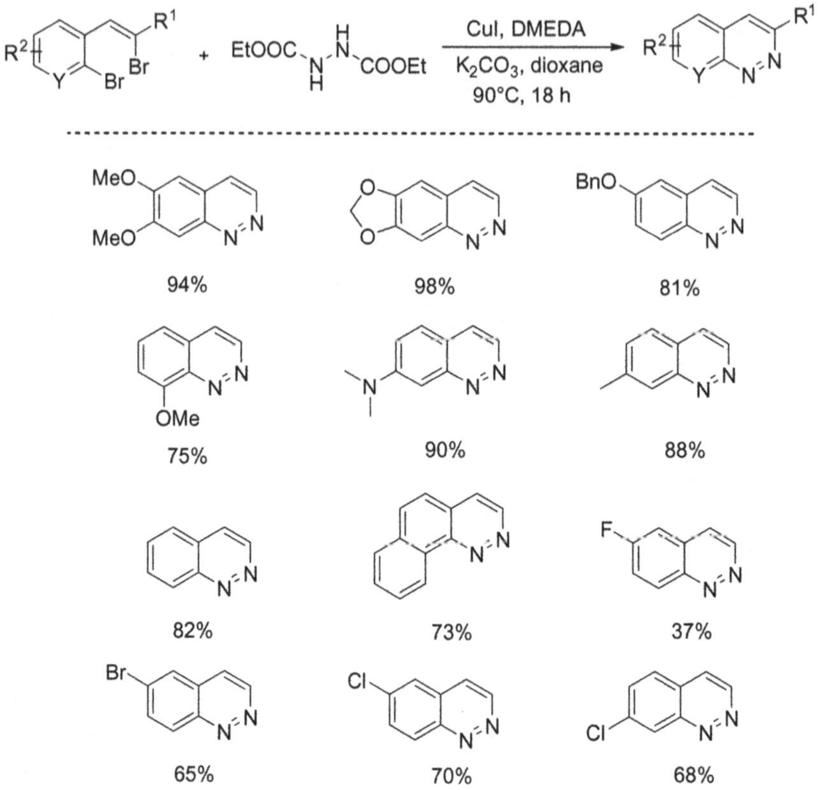

Scheme 4.3 Synthesis of pyridazines.

In 2012, Ge's group developed copper-catalyzed aerobic dehydrogenative cyclization of N-methyl-N-phenylhydrazones to synthesize pyridazines, which went through sequential C_{sp3}-H oxidation, cyclization, and aromatization processes (Scheme 4.4) [4]. There is no apparent electronic or steric effect in this reaction. The substrates with both electron-donating or electron-withdrawing substituents (R^1) on either the para, meta, and ortho positions could work well. And good to high yields of products were obtained. The meta-OMe-, Me-, or

Br-substituted substrates gave a mixture of para- and orthosubstituted products with a preference for the *p*-substituted products, whereas substrates bearing the more hindered *i*Pr group and the electron-withdrawing CN group provided only the *p*-substituted products. In contrast, there is an electronic effect resulting from substituents (R^2) on the other phenyl ring. Generally, electron-donating groups on this ring provide higher yields than those with electron-withdrawing groups.

Scheme 4.4

A 50 mL Schlenk tube was charged with N-methyl-N-phenylhydrazones (1, 0.3 mmol), $CuSO_4$ (1.0 mg, 0.0045 mmol), CuI (4.2 mg, 0.0225 mmol), Py (84.4 mL, 1.05 mmol), and DMF (2.7 mL). Then a solution of CF_3SO_3H (26.5 mL, 0.3 mmol) in DMF (0.3 mL) was slowly added. The tube was evacuated and filled with 1 atm O_2, and stirred rigorously at 110°C (unless otherwise noted) for 14−48 h. After removal of the solvent, the residue was purified by flash chromatography on silica gel.

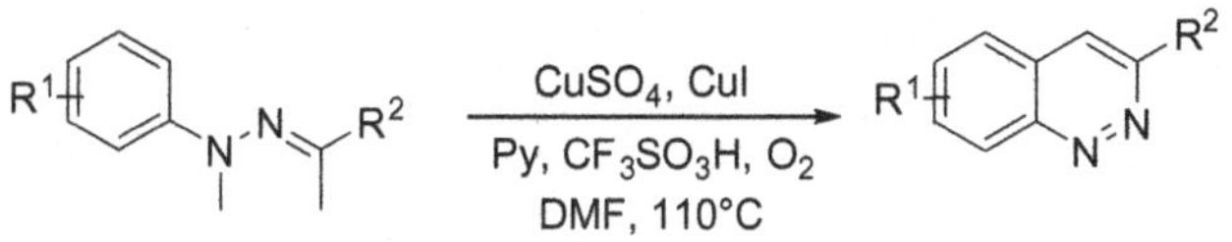

Scheme 4.4 Copper-catalyzed synthesis of pyridazines.

REFERENCES

[1] Pridmore, S. J.; Slatford, P. A.; Taylor, J. E.; Whittlesey, M. K.; Williams, J. M. J. *Tetrahedron* **2009**, *65*, 8981−8986.

[2] Zhu, C.; Yamane, M. *Tetrahedron* **2011**, *67*, 4933−4938.

[3] Ball, C. J.; Gilmore, J.; Willis, M. C. *Angew. Chem. Int. Ed.* **2012**, *51*, 5718−5722.

[4] Zhang, G.; Miao, J.; Zhao, Y.; Ge, H. *Angew. Chem. Int. Ed.* **2012**, *51*, 8318−8321.

CHAPTER 5

Synthesis of Triazine

In 1984, Vollhardt and coworkers reported a partially intramolecular synthesis of triazines by using the Fe-catalyzed cyclotrimerization of adiponitrile with nitriles under 220°C (Scheme 5.1) [1]. The reaction tolerated alkyl, benzyl, and phenyl nitriles, providing triazines in moderate to good yields.

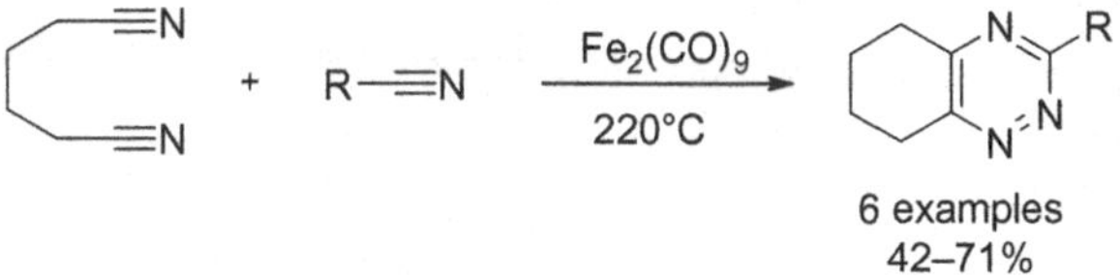

N
N
R
N
$Fe_2(CO)_9$
220°C
N
R
N
N
6 examples
42–71%

Scheme 5.1 Fe-catalyzed synthesis of triazines.

In 2009, a simple, new three-step sequence for the conversion of hydrazides into triazines was reported by Moody and coworkers (Scheme 5.2) [2]. The reaction were through N–H insertion by a copper carbene intermediate derived from α-diazo-β-ketoesters into the hydrazide, reaction with ammonium acetate to give 1,2,4-triazines. And a range of substituted benzhydrazides were converted into 1,2,4-triazine. In this reaction, copper(II) acetate was the catalyst of choice when they used benzhydrazide as a model substrate with methyl 2-diazo-3-oxobutanoate as the diazocarbonyl component. The reaction was carried out in a microwave reactor. The product was then treated with ammonium acetate in acetic acid to give, after chromatography, the 1,2,4-triazine in modest yield. Different triazines were obtained using this method.

Scheme 5.2

To a microwave tube were added the hydrazide (c.1.0 mmol), copper(II) acetate (6 mg, 0.035 mmol), methyl 2-diazo-3-oxobutanoate (100 mg, 0.7 mmol) in dichloromethane (1 mL). The mixture was subject to microwave irradiation (sealed tube, 200 W) at 80°C for 10 min. The mixture was filtered through silica gel, washing through with 50% ethyl acetate in

Transition Metal Catalyzed Pyrimidine, Pyrazine, Pyridazine and Triazine Synthesis.
DOI: http://dx.doi.org/10.1016/B978-0-12-809378-8.00005-5

cyclohexane, and the filtrate evaporated. Ammonium acetate (270 mg, 3.5 mmol) and acetic acid (1 mL) were added to the residue, and the mixture was subjected to microwave irradiation (50 W) at 110°C for 5 min. The mixture was neutralized by addition of saturated sodium hydrogen carbonate solution (5 mL), and the product was extracted by ethyl acetate (2 × 10 mL). The combined extracts were washed with saturated sodium hydrogen carbonate solution (10 mL), water (10 mL), and brine (10 mL), dried over MgSO$_4$ and concentrated in vacuo. The product was purified by chromatography.

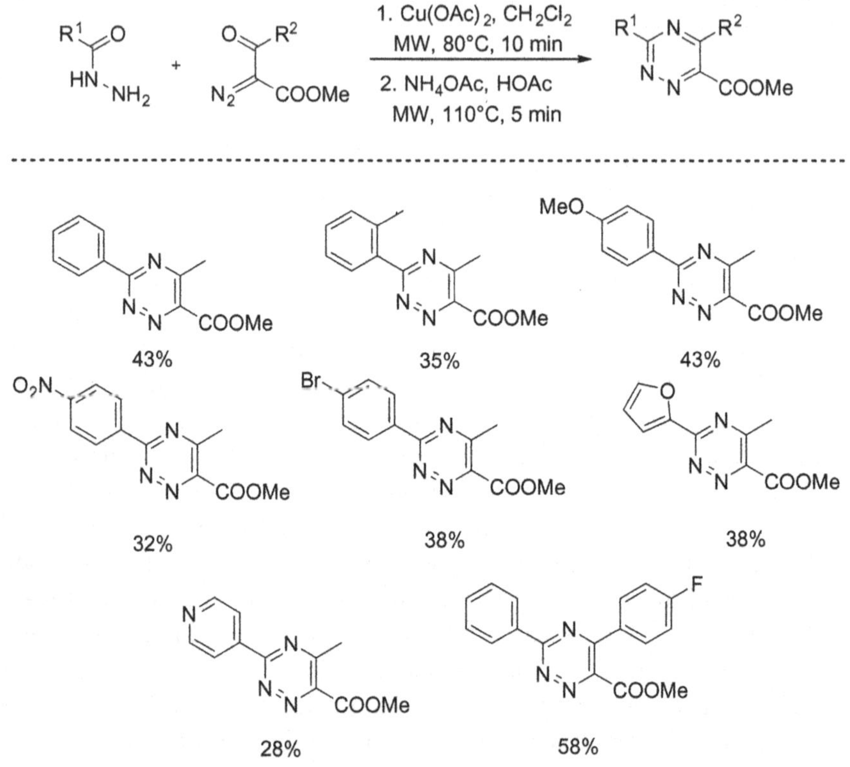

Scheme 5.2 Synthesis of triazines.

In 2010, Shinde reported the one-pot synthesis of a series of 1,2,4-triazines. In this article, they used 1-(1-substituted piperidin-4-yl)-1H-1,2,3-triazole-4-carbohydrazide and benzyl as the substrates, ammonium acetate and ZrOCl$_2$.8H$_2$O as a catalyst in the mixed solvents of ethanol and water (Scheme 5.3), which gave the products in the range of

87–94% yields [3]. In particular, the synthesized compounds were tested for in vitro antifungal activity.

Scheme 5.3

To a stirred solution of different hydrazides (10 mmol) in 15 mL ethanol: water (1:2) was added benzil (10 mmol), ammonium acetate (40 mmol), and $ZrOCl_2 \cdot 8H_2O$ (1 mmol). The reaction mixture was heated at 100°C for 100–120 min. After completion (monitored by TLC), solvent was removed in vacuo and distilled water was added to the residue and then extracted with three portions of ethyl acetate. The combined organic layers were dried (Na_2SO_4) and concentrated in vacuo to get the desired compounds. The crude products obtained were purified by recrystallization using ethanol.

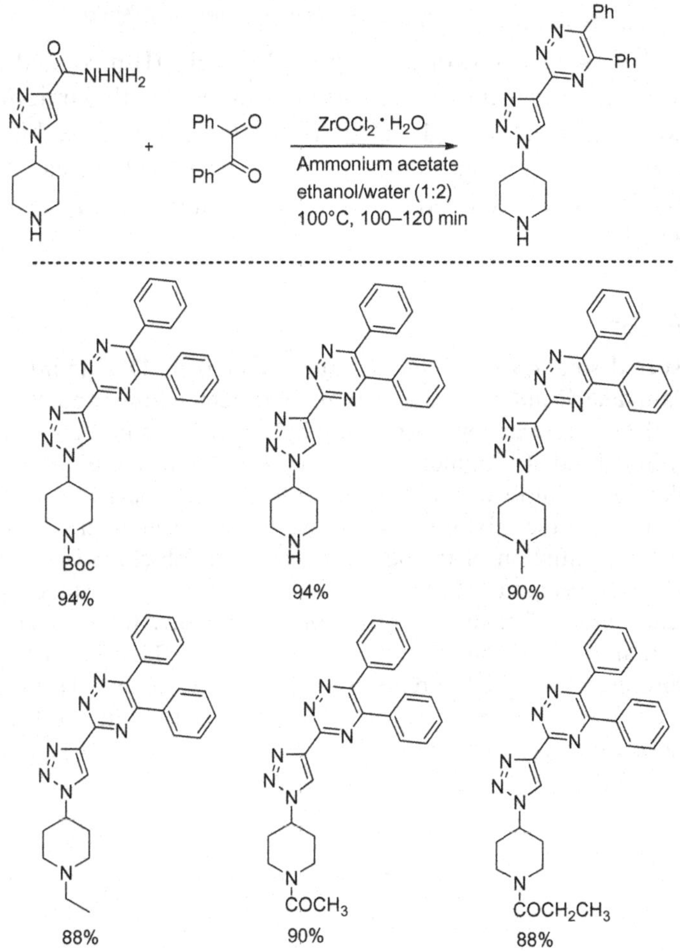

Scheme 5.3 One-pot synthesis of 1,2,4-triazines.

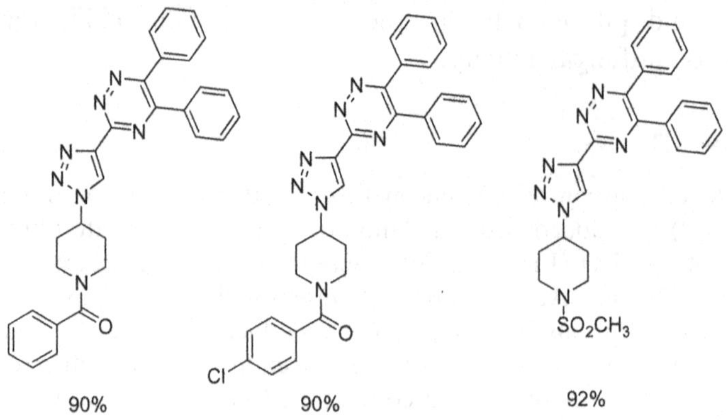

90%   90%   92%

Scheme 5.3 (Continued)

In 2010, Vasu and coworkers reported the Hg(II)-mediated one-pot synthesis of 1,3,5-triazines from isothiocyanates, diethylamidines, and carbamidines(Scheme 5.4) [4]. Initially, the reaction of isothiocyanate with amidine produced the corresponding amidinothiourea. A subsequent Hg(II)-mediated condensation of the latter with carbamidine furnished triazines.

Scheme 5.4

To a stirred solution of isothiocyanate (1 mmol) in THF (5 mL), N,N-diethylamidine (1 mmol) was added at room temperature and stirred for 2 h. To the stirred reaction mixture triethylamine (15 mmol) and carbamidine hydrochloride (1.2 mmol) were added at room temperature. Mercury (II) chloride (1 mmol) was added slowly to the reaction mixture (slightly exothermic) and the mixture was allowed to stir for another 4 h at room temperature. Initiation of the desulfurization was observed by formation of black HgS precipitate. Progress of the reaction was monitored by TLC using ethyl acetate:hexane (3:7) as a mobile phase. After completion of the reaction, the reaction mixture was diluted with THF (5 mL) and filtered through a celite bed to remove the black precipitate of HgS. The filtrate was concentrated in vacuo and purified by silica gel column chromatography using triazines.

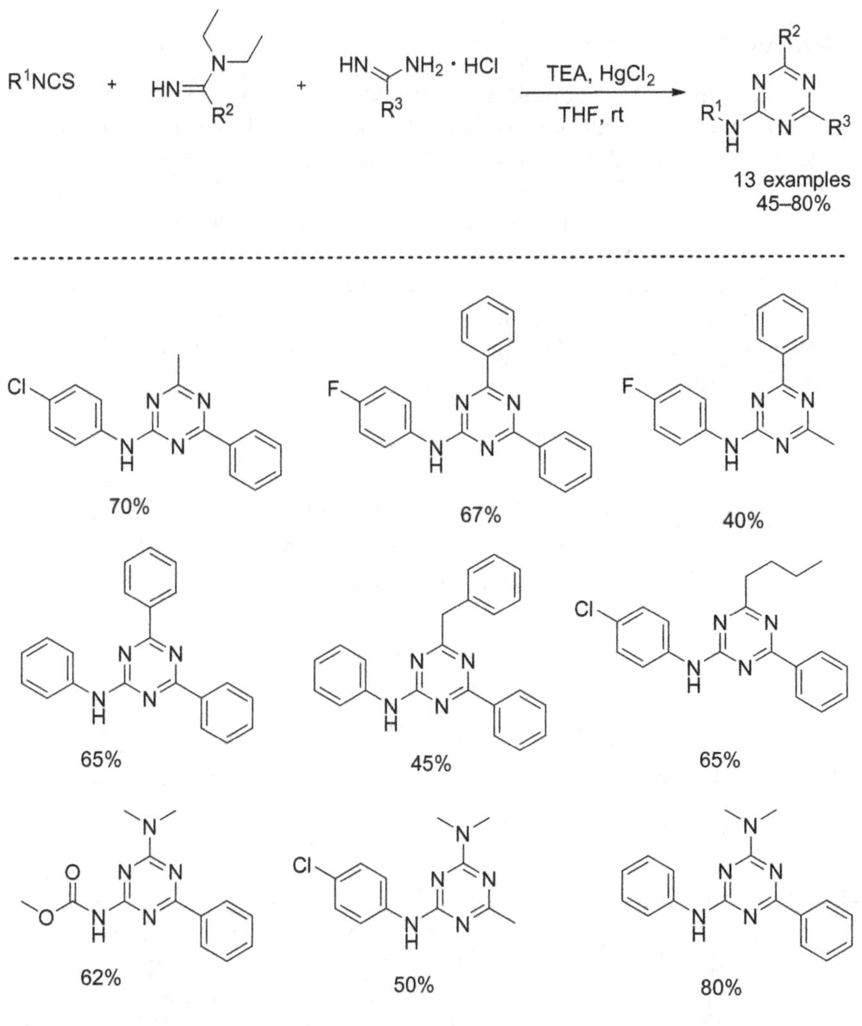

Scheme 5.4 Hg(II)-mediated one-pot synthesis of 1,3,5-triazines.

In 2012, Batra developed a one-step synthesis of 1,3,5-triazines through a copper-catalyzed cascade reaction, which gave products in moderate to good yields (Scheme 5.5) [5]. First, they used indole-2-carbaldehyde and acetamidine hydrochloride or benzamidine hydrochloride as the substrates. And 1,3,5-triazines were obtained in good yields. The reaction worked smoothly though the substrates indole-2-carbaldehydes were with electron-donating or electron-withdrawing groups. Then benzaldehyde with electron-donating or electron-withdrawing groups were chosen as the substrates. The reactions proceeded smoothly to furnish the respective substituted 1,3,5-triazines in good to excellent isolated yields.

Scheme 5.5

To a solution of 3-iodoindole-2-carbaldehyde (0.3 g, 1.1 mmol) and acetamidine hydrochloride (0.21 g, 2.2 mmol) in DMSO (2 mL) was added Cs_2CO_3 (0.72 g, 2.2 mmol), and the reaction mixture was heated at 90°C for 12 h in a sealed tube under nitrogen. Thereafter, water (50 mL) and ethyl acetate (25 mL) were added, and the reaction mass and the organic layers were separated. The aqueous layer was extracted with ethyl acetate (2 × 20 mL). The combined organic layers were washed with brine (35 mL), dried with anhydrous Na_2SO_4, and concentrated under vacuum. The product was purified by chromatography (EtOAc/hexanes, 1:4).

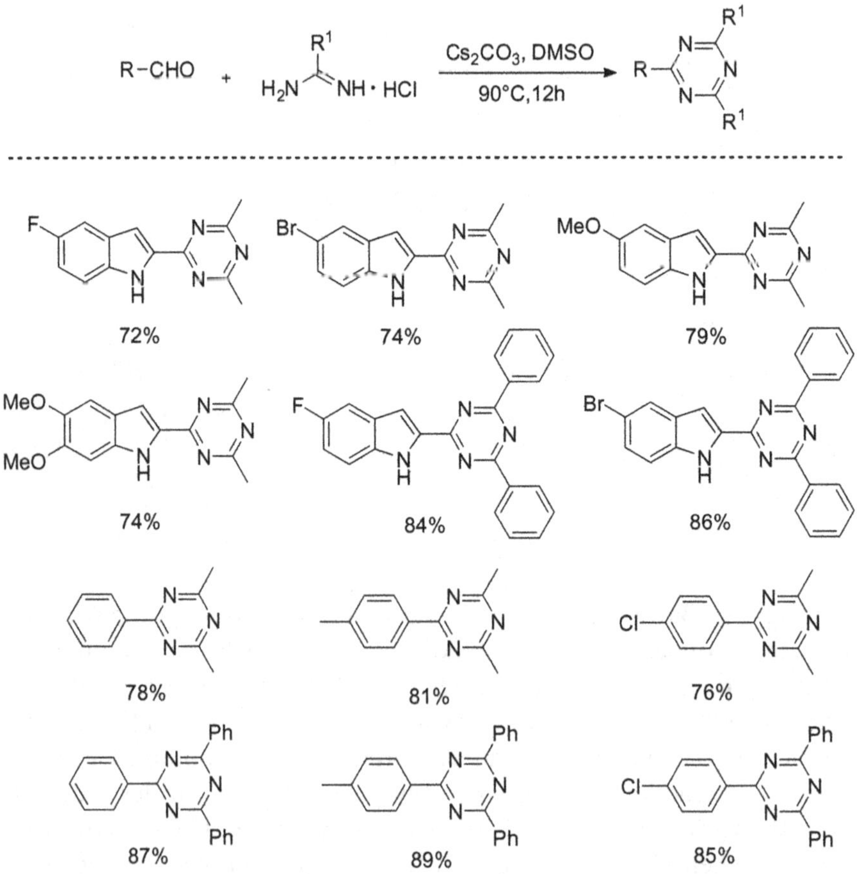

Scheme 5.5 Copper-catalyzed synthesis of polyfunctionalized 1,3,5-triazines.

In 2014, Zhang's group reported an efficient ruthenium-catalyzed dehydrogenative synthesis of 2,4,6-triaryl-1,3,5-triazines from aryl methanols and amidines (Scheme 5.6) [6]. In this reaction, they used [RuCl$_2$(p-cymene)]$_2$/Cs$_2$CO$_3$ as the catalyst system. Due to the inherent stability of alcohols in contrast with aldehydes, their synthetic protocol was adaptable to a broad substrate scope. The reactions of aryl and alkyl amidines in combination with different substituted benzylic alcohols were examined, which proceeded smoothly and furnished desired products in good to excellent yields. Besides, halogen as well as electron-donating groups were apparently well tolerated. Even benzylic alcohols having a strong electron-withdrawing group (-NO$_2$) with different substitution patterns could also be transformed into the corresponding products in an efficient manner, and the metasubstituted one gave the product in a higher yield; it was proposed that the metasubstituted substrate was relatively more stable than those of para- and orthosubstituted ones under the standard conditions. While using the same amidine, the substituents possessing different electron properties on the aryl ring of benzylic alcohols slightly influenced the product yields. In general, the electron-rich substrate such as 4-methoxy benzylic alcohol gave the products in relatively higher yields, which could also be assigned to its inherent stability in comparison with other benzylic alcohols. Then they used some representative heteroaryl substituted methanols such as pyridyl methanols, furyl methanol, and thiazol-2-ol as the substrates. Gratifyingly, all the reactions also proceeded well.

Scheme 5.6

To a solution of benzyl alcohol (0.162 g, 1.5 mmol) and benzamidine hydrochloride (0.156 g, 1 mmol) in DMSO (1 mL) were added [RuCl$_2$-(p-cymene)]$_2$ (0.015 mmol, 4.5 mg) and Cs$_2$CO$_3$ (0.325 g, 1 mmol). The reaction mixture was heated at 110°C for 16 h in a sealed tube without inserting any gas protection. Afterwards, water (10 mL) and dichloromethane (20 mL) were added, the layers were separated, and then the aqueous layer was extracted with dichloromethane (2 × 10 mL). The combined organic layers were dried with anhydrous Na$_2$SO$_4$, and concentrated under vacuum. The residue was directly purified by flash chromatography on silica.

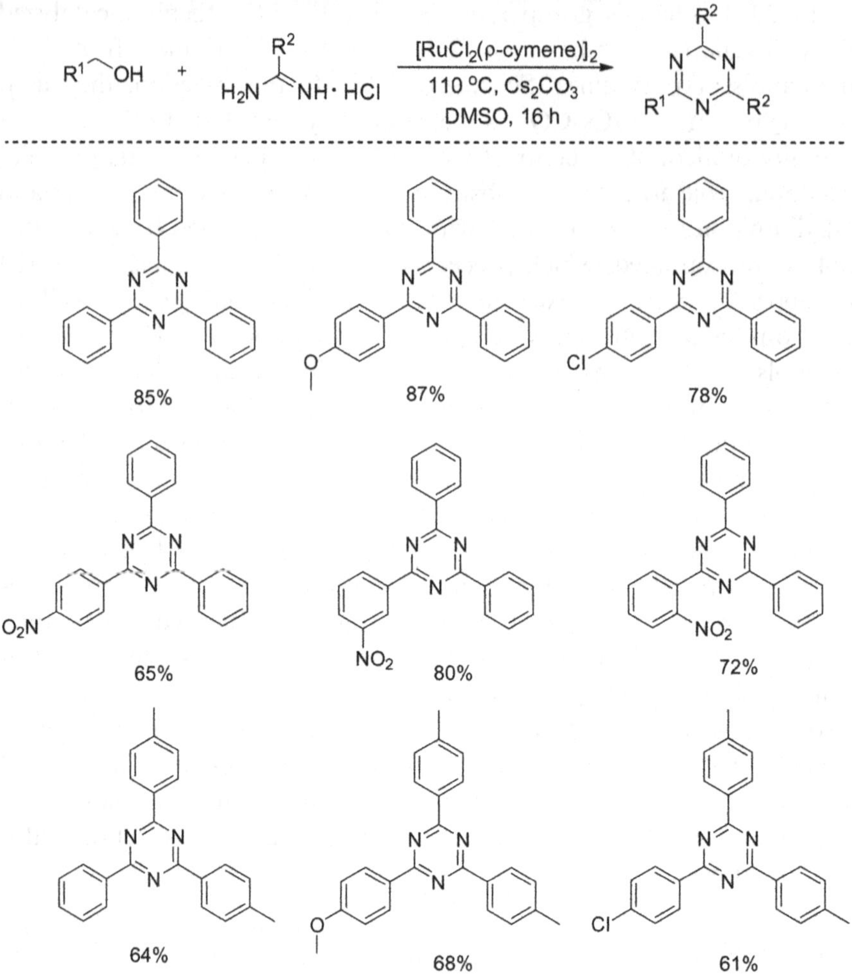

Scheme 5.6 Ruthenium-catalyzed synthesis of 1,3,5-triazines.

In 2014, Li and coworkers developed a novel straightforward synthesis of 2,4-disubstituted-1,3,5-triazines via aerobic copper-catalyzed cyclization of amidines with DMF (Scheme 5.7) [7]. They focused on the synthesis of symmetrical 2,4-diaryl-1,3,5-triazines by using a variety of aryl amidines. All of the reactions proceeded smoothly and furnished the desired products in moderate to good yields. It was found that electron-donating groups (-Me, -OMe) containing amidines afforded the products in higher yields than the electron-deficient ones. However, the homocoupling of acetamidine failed to give the desired product. It was conceivable that the alkyl amidines disfavor the formation of essential intermediates owing to the lack of aryl stabilizing groups. Unsymmetrical 2,4-disubstituted-1,3,5-triazines with the synthetic protocol were obtained as well. By employing different combinations of aryl amidines, all of the cross-coupling reactions underwent efficient cyclization to afford the desired products in moderate to good yields. The electron-rich amidines also gave the products in relatively higher yields than the electron-poor ones.

Scheme 5.7

The mixture of amidine 1 (0.5 mmol), amidine 1' (2.0 mmol), CuI (20 mol%), pyridine (40 mol%) and Et_3N (2 mmol) in DMF (2.0 ml) was stirred at 90°C for 14 h under 1 atm of O_2 atmosphere. Then, the reaction mixture was cooled down to room temperature. EtOAc (20 mL × 3) and H_2O (20 mL × 3) were added successively to extract the crude product, and the combined organic layer was dried with anhydrous Na_2SO_4 and then evaporated under vacuum to remove the organic solvent. Then the resulting residue was directly purified by flash chromatography.

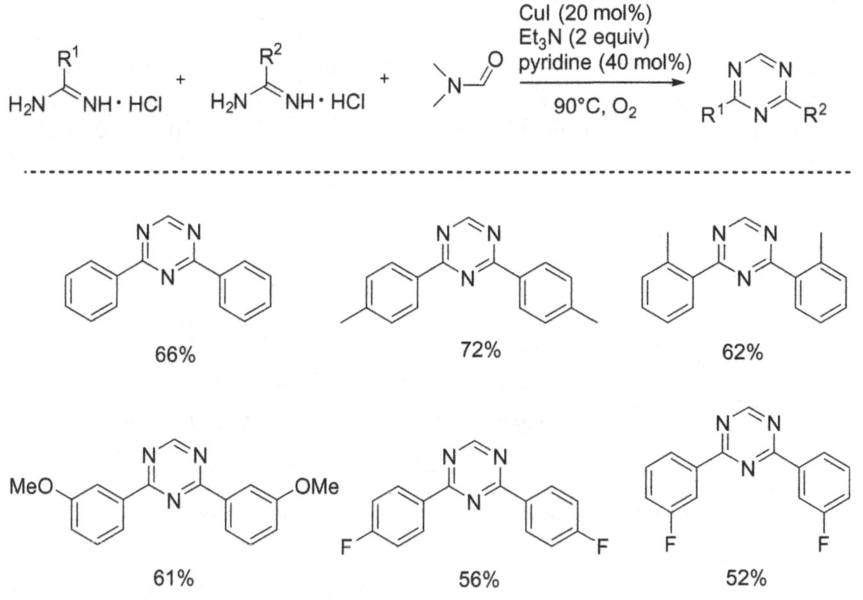

Scheme 5.7 Copper-catalyzed synthesis of 2,4-disubstituted-1,3,5-triazines.

In 2015, Li and Jiang reported copper-catalyzed oxidative C($sp3$)-H functionalization for facile synthesis of 1,2,4-triazoles and 1,3,5-triazines from amidines, which used copper as the catalyst, K_3PO_4 as the base, and O_2 as the oxidant (Scheme 5.8) [8]. A wide range of the two three-nitrogen-containing heterocycles could be readily accessed by using three different reaction partners: trialkylamines, DMSO, and DMF. Generally, arylamidines bearing either electron-donating or electron-withdrawing groups were all efficient substrates, reacting with TEA to provide the desired products in good yields. Fortunately, by using two different amidines with high polarity difference, a range of cross-cyclized triazines were isolated in moderate yields. And a series of arylamidines were subjected to the DMSO- and DMF-based systems, respectively. It was found that various functional groups on the benzene ring, such as methyl, fluoro, chloro, bromo, trifluoromethyl, and methoxy, were well tolerated in both systems. The desired three-nitrogen heterocycles were prepared in moderate to good yields.

Scheme 5.8

A mixture of amidine, Cu(OAc)$_2$ (4.5 mg, 10 mol %), K$_3$PO$_4$ (106 mg, 2 equiv), and toluene (1 mL) in a test tube (10 mL) equipped with a magnetic stirring bar. The mixture was stirred under 1 atm O$_2$ atmosphere at 100°C for 10 h. After the reaction was completed, 10 mL ethyl acetate (3 × 10 mL) was added into the tube. The combined organic layers were washed with brine to neutral, dried over MgSO$_4$, and concentrated in vacuum. Purification of the residue on a preparative TLC afforded the product.

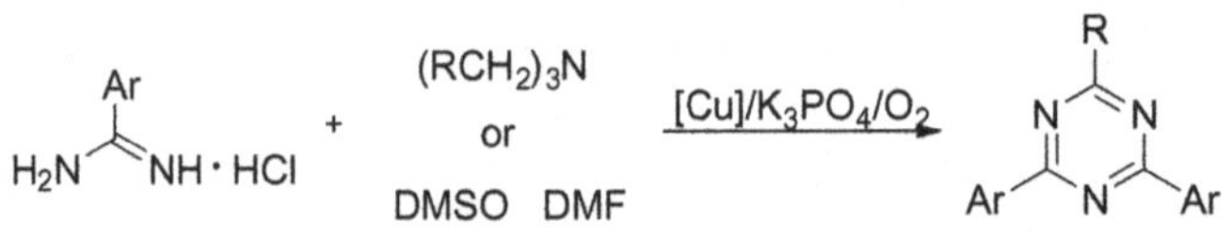

Scheme 5.8 Copper-catalyzed synthesis of oxidative C(sp3)-H functionalization 1,3,5-triazines.

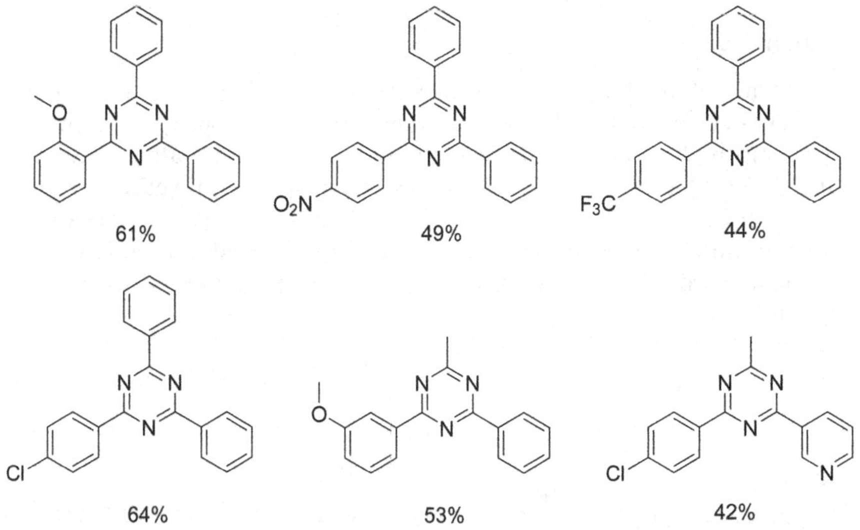

61%   49%   44%

64%   53%   42%

Scheme 5.8 (Continued)

In 2015, Zhang's group demonstrated the synthesis of 1,3,5-triazine derivatives directly by oxidative coupling of amidine hydrochlorides and alcohols in air with $Cu(OAc)_2$ as a catalyst (Scheme 5.9) [9]. A variety of primary alcohols and amidine hydrochlorides with different substituents were exploited as starting reactants for the synthesis of 1,3,5-triazine derivatives. Good results were obtained as primary aliphatic alcohols such as methanol, ethanol, n-butyl alcohol, and isobutanol and were tested. Several functional groups, such as methoxy, chloro, fluoro, and nitro were well tolerated. No significant substitution effect was observed on aromatic alcohols, excellent yields were obtained for alcohols with both electron-donating and electron-withdrawing substituents on the para-position of aryl rings. Relatively low yields were obtained when alcohols were substituted with electron-withdrawing substituents such as Br and NO_2. Besides, the steric hindrance had an obvious effect on the reaction. When methyl-substituted benzyl alcohols were used, para- and metasubstituted substances gave higher yields than those with ortho substituents.

Scheme 5.9

A mixture of amidine, $Cu(OAc)_2$ (4.5 mg, 10 mol %), K_3PO_4 (106 mg, 2 equiv), and toluene (1 mL) in a test tube (10 mL) equipped with a magnetic stirring bar. The mixture was stirred under 1 atm O_2 atmosphere at 100°C for 10 h. After the reaction was completed, 10 mL ethyl acetate (3 × 10 mL) was added into the tube. The combined organic layers were washed with brine to neutral, dried over $MgSO_4$, and concentrated in vacuum. Purification of the residue on a preparative TLC afforded the product.

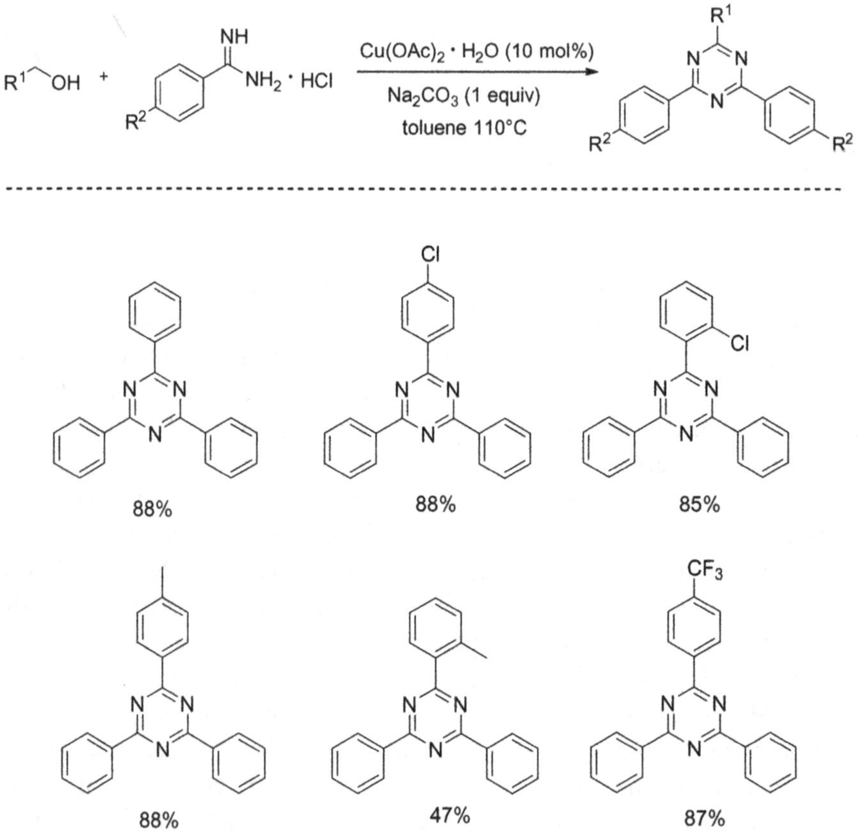

Scheme 5.9 Copper-catalyzed synthesis of 1,3,5-triazines.

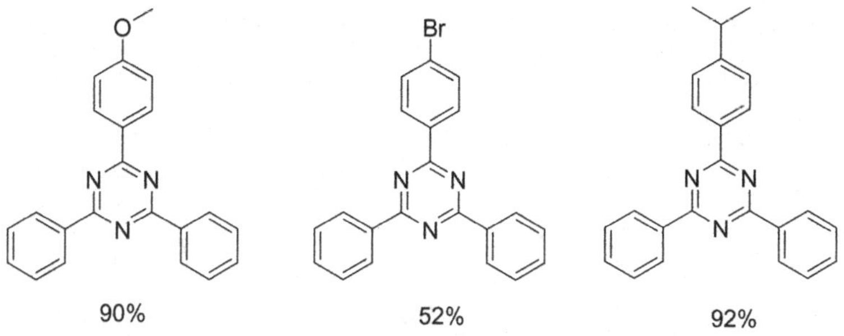

90%   52%   92%

Scheme 5.9 (Continued)

Recently, Krasavin and coworkers developed the Zn-catalyzed one-pot method to 1,2,4-triazines via a Zn-catalyzed hydrohydrazination–cyclodehydration–oxidation sequence involving propargylamides and BocNHNH$_2$, in moderate to good yields (Scheme 5.10) [10]. A range of triazines were obtained using this method. Various propargylamines were used in this reaction and yielded respective 1,2,4-triazines in moderate to good yields. And many of the compounds obtained in this work could be considered as suitable tools for fragment-based drug discovery.

Scheme 5.10

The respective propargylamide (1.26 mmol) in dry toluene (15 mL) was treated with BocNHNH2 (6, 1.26 mmol) and Zn(OTf)$_2$ (0.32 mmol). The resulting mixture was heated under reflux in the atmosphere of argon for 4 h and then cooled down to room temperature. Aqueous solution (10 mL) containing K$_3$[Fe(CN)$_6$] (0.62 g, 1.89 mmol) and NaOH (0.13 g, 3.15 mmol) was added. The resulting biphasic mixture was vigorously stirred overnight. The organic phase was separated, dried over anhydrous Na$_2$SO$_4$, filtered, and concentrated in vacuo. The residue was purified by column chromatography on silica gel.

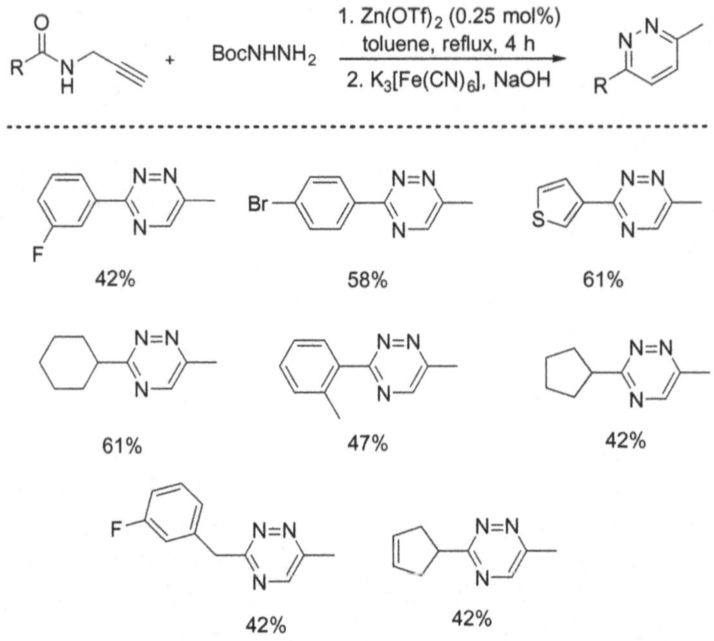

Scheme 5.10 Zn-catalyzed synthesis of 1,2,4-triazines.

In 2016, Li and coworkers reported an efficient iridium-catalyzed dehydrogenation, ring-closure reaction, in which [Cp*IrI₂]₂/Xantphos were proved to be the most efficient catalyst for the synthesis of 1,3,5-triazines from stable aryl-substituted alcohols and amidines (Scheme 5.11) [11]. Generally, the reaction had high substituent tolerance. The substituents on pyridyl methanol substrates had little impact on formation of the desired products. The effect of substituents on the aromatic ring of alcohols was studied. Benzamidine hydrochloride with benzyl alcohols having electron-donating or electron-withdrawing substituents produced moderate to high yields. Furyl- and thiazolyl-heterocyclic alcohols reacted under the optimized conditions with satisfactory yields. Besides, the reaction of substituted benzamidines with aryl methanols also gave the corresponding 2,4,6-triaryl-1,3,5-triazines in high yields. Substituents with different electronic properties on the aryl ring of benzamidines significantly affected the reaction. Benzamidines possessing electron-poor groups led to corresponding products in higher yields than the electron-donating ones.

Scheme 5.11

[Cp*IrI$_2$]$_2$ (1 mol %, 0.01 mmol), xantphos (2 mol %, 0.02 mmol), and dioxane (2 mL) were stirred briefly in a Schlenk tube at room temperature. Subsequently, 3-pyridinemethanol (2.0 mmol), benzamidine hydrochloride (1.0 mmol), and cesium carbonate (1.0 mmol) were added. The mixture was heated under 110°C for 20 h and then cooled down to room temperature. The resulting solution was purified by column chromatography.

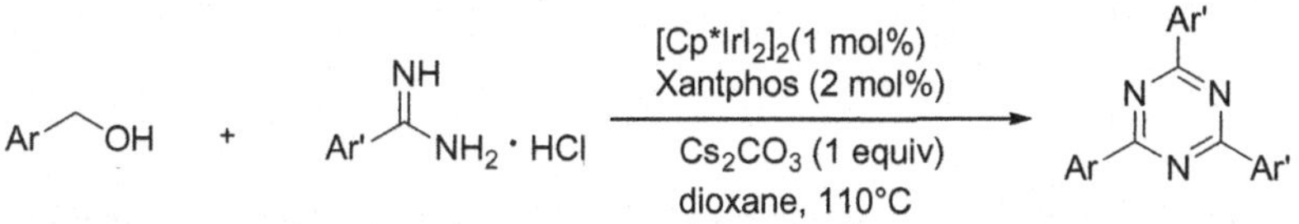

Scheme 5.11 Iridium-catalyzed synthesis of 1,3,5-triazines.

REFERENCES

[1] Gesing, E. R. F.; Groth, U.; Vollhardt, K. P. C. *Synthesis* **1984**, *4*, 351–353.

[2] Shi, B.; Lewis, W.; Campbell, I. B.; Moody, C. J. *Org. Lett.* **2009**, *11*, 3686–3688.

[3] Sangshetti, J. N.; Shinde, D. B. *Bioorg. Med. Chem. Lett.* **2010**, *20*, 742–745.

[4] Kaila, J. C.; Baraiya, A. B.; Pandya, A. N.; Jalani, H. B.; Sudarsanam, V.; Vasu, K. K. *Tetrahedron Lett* **2010**, *51*, 1486–1489.

[5] Biswas, S.; Batra, S. *Eur. J. Org. Chem.* **2012**, 3492–3499.

[6] Xie, F.; Chen, M.; Wang, X.; Jiang, H.; Zhang, M. *Org. Biomol. Chem.* **2014**, *12*, 2761–2768.

[7] Xu, X.; Zhang, M.; Jiang, H.; Zheng, J.; Li, Y. *Org. Lett.* **2014**, *16*, 3540–3543.

[8] Huang, H.; Guo, W.; Wu, W.; Li, C.; Jiang, H. *Org. Lett.* **2015**, *17*, 2894–2897.

[9] You, Q.; Wang, F.; Wu, C.; Shi, T.; Min, D.; Chen, H., et al. *Org. Biomol. Chem.* **2015**, *13*, 6723–6727.

[10] a. Lukin, A.; Vedekhina, T.; Tovpeko, D.; Zhuriloa, N.; Krasavin, M. *RSC Adv.* **2016**, *6*, 57956–57959.
b. Wu, X. F. *Chem. Rec.* **2015**, *15*, 949–963.
c. Wu, X.-F.; Neumann, H. *Adv. Synth. Catal.* **2012**, *354*, 3141–3160.
d. Wu, X. F. *Chem. Asian J.* **2012**, *7*, 2502–2509.

[11] Shi, G.; He, F.; Che, Y.; Nia, C.; Li, Y. *Russ. J. Gen. Chem.* **2016**, *86*, 380–386.

CHAPTER 6

Summary

The main achievements of transition metal-catalyzed synthesis of pyrimidine, pyrazine, pyridazine and triazine have been summarized and discussed. The whole manuscript is divided by the type of the reactions. Hopefully, it will be useful for our community.

As expected, there may be mistakes. We are here asking for your forgiveness for these and also for the possible missed literature.

Transition Metal Catalyzed Pyrimidine, Pyrazine, Pyridazine and Triazine Synthesis.
DOI: http://dx.doi.org/10.1016/B978-0-12-809378-8.00006-7

Printed in the United States
By Bookmasters